Testing and Evaluation of

Standoff Chemical Agent Detectors

Committee on Testing and Evaluation of Standoff Chemical Agent Detectors

Board on Chemical Sciences and Technology

Division on Earth and Life Studies

NATIONAL RESEARCH COUNCIL
OF THE NATIONAL ACADEMIES

THE NATIONAL ACADEMIES PRESS
Washington, D.C.
www.nap.edu

THE NATIONAL ACADEMIES PRESS 500 Fifth Street, N.W. Washington, DC 20001

NOTICE: The project that is the subject of this report was approved by the Governing Board of the National Research Council, whose members are drawn from the councils of the National Academy of Sciences, the National Academy of Engineering, and the Institute of Medicine. The members of the committee responsible for the report were chosen for their special competences and with regard for appropriate balance.

This study was supported by Contract No. DTRA 01-02-C-0091 between the Department of Defense and the National Research Council. Any opinions, findings, conclusions, or recommendations expressed in this publication are those of the authors and do not necessarily reflect the views of the organization or agencies that provide support for the project.

International Standard Book Number 0-309-08740-6 (Book)
International Standard Book Number 0-309-50712-X (PDF)

Additional copies of this report are available from:

The National Academies Press
500 5th Street, N.W.
Lockbox 285
Washington, DC 20055
800-624-6242
202-334-3313 (in the Washington Metropolitan Area)
http://www.nap.edu

Printed in the United States of America

THE NATIONAL ACADEMIES

Advisers to the Nation on Science, Engineering, and Medicine

The **National Academy of Sciences** is a private, nonprofit, self-perpetuating society of distinguished scholars engaged in scientific and engineering research, dedicated to the furtherance of science and technology and to their use for the general welfare. Upon the authority of the charter granted to it by the Congress in 1863, the Academy has a mandate that requires it to advise the federal government on scientific and technical matters. Dr. Bruce M. Alberts is president of the National Academy of Sciences.

The **National Academy of Engineering** was established in 1964, under the charter of the National Academy of Sciences, as a parallel organization of outstanding engineers. It is autonomous in its administration and in the selection of its members, sharing with the National Academy of Sciences the responsibility for advising the federal government. The National Academy of Engineering also sponsors engineering programs aimed at meeting national needs, encourages education and research, and recognizes the superior achievements of engineers. Dr. Wm. A. Wulf is president of the National Academy of Engineering.

The **Institute of Medicine** was established in 1970 by the National Academy of Sciences to secure the services of eminent members of appropriate professions in the examination of policy matters pertaining to the health of the public. The Institute acts under the responsibility given to the National Academy of Sciences by its congressional charter to be an adviser to the federal government and, upon its own initiative, to identify issues of medical care, research, and education. Dr. Harvey V. Fineberg is president of the Institute of Medicine.

The **National Research Council** was organized by the National Academy of Sciences in 1916 to associate the broad community of science and technology with the Academy's purposes of furthering knowledge and advising the federal government. Functioning in accordance with general policies determined by the Academy, the Council has become the principal operating agency of both the National Academy of Sciences and the National Academy of Engineering in providing services to the government, the public, and the scientific and engineering communities. The Council is administered jointly by both Academies and the Institute of Medicine. Dr. Bruce M. Alberts and Dr. Wm. A. Wulf are chair and vice chair, respectively, of the National Research Council.

www.national-academies.org

COMMITTEE ON TESTING AND EVALUATION OF STANDOFF CHEMICAL AGENT DETECTORS

EDWIN P. PRZYBYLOWICZ, Eastman Kodak Company (Retired), Chair
EDWARD V. BROWELL, NASA Langley Research Center
D. BRUCE CHASE, E. I. Dupont de Nemours & Company
JAMES A. DE HASETH, University of Georgia
RICHARD C. FLAGAN, California Institute of Technology
PETER R. GRIFFITHS, University of Idaho
DAVID W. LAYTON, Lawrence Livermore National Laboratory
THOMAS A. REICHARDT, Sandia National Laboratories
LORENZ R. RHOMBERG, Gradient Corporation

Staff

CHRISTOPHER K. MURPHY, Program Officer
SYBIL A. PAIGE, Administrative Associate
DOUGLAS J. RABER, Senior Scholar
DAVID C. RASMUSSEN, Program Assistant
DOROTHY ZOLANDZ, Director, Board on Chemical Sciences and Technology

Acknowledgment of Reviewers

This report has been reviewed in draft form by individuals chosen for their diverse perspectives and technical expertise, in accordance with procedures approved by the National Research Council's Report Review Committee. The purpose of this independent review is to provide candid and critical comments that will assist the institution in making the published report as sound as possible and to ensure that the report meets institutional standards for objectivity, evidence, and responsiveness to the study charge. The review comments and draft manuscript remain confidential to protect the integrity of the deliberative process. We wish to thank the following individuals for their review of this report:

Steven D. Brown, University of Delaware
Augustus W. Fountain III, United States Military Academy
Stanley Kaplan, Bayesian Systems, Incorporated
W. Carl Lineberger, University of Colorado
Alan B. Rothman, Argonne National Laboratory
Lynn M. Russell, Princeton University
Panagiotis G. Smirniotis, University of Cincinnati
J.D. Tate, Dow Chemical Company

Although the reviewers listed above have provided many constructive comments and suggestions, they were not asked to endorse the conclusions or recommendations nor did they see the final draft of the report before its release. The review of this report was overseen by Lewis M. Branscomb, Harvard University, and Royce W. Murray, University of North Carolina at Chapel Hill. Appointed by the National Research Council, they were responsible for making certain that an independent examination of this report was carried out in accordance with institutional procedures and that all review comments were carefully considered. Responsibility for the final content of this report rests entirely with the authoring committee and the institution.

Preface

The present study was requested by the Department of Defense (DoD) as a result of questions raised regarding the conclusions of an earlier study carried out by the Battelle Memorial Institute (BMI). The DoD asked that the National Research Council provide a "second opinion" for how to test and evaluate standoff chemical warfare agent (CWA) detectors. Because of the possible imminent need to deploy such detectors, this NRC study was conducted over a short time frame—four months. The committee's Statement of Work, in its narrow but important focus, reflects the urgent nature of this study.[1]

Despite the complex technical issues addressed by the committee, this study and report contribute but a small piece to the very large and multifaceted challenge of detecting CWAs in the environment. There is a large body of information already in existence on this broad topic within the DoD and other agencies that have addressed not only the CWA topic itself but also related subjects such as delivery systems, terrorism activities, and strategies for the use of such agents. In related areas, there are meteorological and space studies that can provide necessary basic information important to modeling the behavior of CWAs in the environment. This body of knowledge can be valuable to understanding the field deployment of standoff detectors not only of the present design but also of future detectors that may be based on alternative technologies. These are areas that the committee determined were outside its purview based on its Statement of Work.

The committee believes that it has provided in this report a sound technical and practical pathway to the challenge of testing and evaluating standoff detectors for CWAs based on infrared spectroscopy.

[1]The committee's Statement of Work may be found in Appendix A.

Contents

Executive Summary

The detection of chemical warfare agents is a topic that attracts a good deal of attention in the current environment of the war on terrorism. It is well, therefore, to point out at the outset what this report does and does not cover because the broad topic is a critical one with many aspects that need to be addressed urgently. Testing and evaluation of standoff chemical warfare agent (CWA)[1] detectors to certify their suitability for field applications present two major challenges: (1) these detectors cannot readily be tested in a field environment with CWAs without extensive precautions and legal approvals by top government officials and (2) the wide variety of environments in which the detectors might be used may significantly affect the instrument performance. The Defense Threat Reduction Agency (DTRA) of the Department of Defense (DoD) sponsored a study by the Battelle Memorial Institute entitled *The Use of Chemical Agent Simulants in Standoff Detection Testing*. This study issued a final report in October 2001 which stated that "using simulants instead of chemical agents provides an effective means for conducting outdoor operational testing of standoff detection instruments."[2,3] Review of the report by the DoD raised concerns that a testing protocol, absent field testing with CWAs, might not adequately predict the response of these detectors to CWAs in battlefield and homeland defense applications.

DTRA requested that the National Research Council undertake a study of the testing and evaluation of standoff detectors to provide an independent assessment of suitable test protocols that might be useful and reliable. In its Statement of Work, the committee was asked to answer the following:

---

[1]Appendix D contains a list of acronyms and a glossary of terms.

[2]A.R. Blackburn. K.S.K. Chinn, S.D. Fortney, W.A. Ivancic, A.K. Judd, B.D. Lerner, and J. Ontiveros. 2001. *The Use of Chemical Agent Simulants in Standoff Detection Testing*. Battelle Memorial Institute, Columbus, Ohio, CBIAC Contract No. SPO700-00-D-3180. CBIAC Task No. 75.

[3]Simulants are molecules having either spectroscopic or physical properties similar to the corresponding properties of a CWA.

1. What test protocols should be adopted to ensure that standoff CWA detectors will meet operational requirements and why. Consideration should be given to a variety of options to include chamber testing, chamber and simulant testing, and live-agent open-air testing.

2. Identify the challenges associated with executing the recommended protocols.

3. Identify the risks associated with not doing open-air testing using live agents.

4. Using Multi-Service Tactics, Techniques, and Procedures for Risk Management as described in FM 3-100.12, *Air Land Sea Risk Management* dated February 2001, (ref Annex D Risk Assessment Matrix), assess the risk associated with operationally employing standoff CWA detectors that have been tested at three possible levels: (1) baseline—live-agent chamber challenges combined with simulant open-air challenges; (2) baseline plus challenges in a test facility capable of enclosed, long-range live-agent challenges; and (3) baseline plus live-agent open-air challenges. If NAS does not feel qualified to assess the severity component of such a risk assessment, NAS may provide the probability component and defer risk assessment to the DoD.

This report is a narrow and very specific study of the testing and evaluation of infrared-type standoff detectors for CWAs in military situations. It is not a study of various possible methods of detecting chemical warfare agents, nor is it a study of the use of such detectors in broader applications such as homeland security. Rather, it is a study of the best possible way to evaluate such detectors in a realistic way that will ensure that they will detect CWAs, primarily in a wartime scenario.

The committee addressed a number of critical issues:

- Which types of detectors were to be included in any test protocols. The committee determined that two types of detectors, passive infrared and active infrared based on lidar technology, would incorporate technology presently used as well as technology likely to be incorporated in the not-too-distant future.
- The value of field-testing of these detectors with CWAs. The present study provides complex—but necessary—test protocols that eliminate the need for field testing with CWAs.
- The risk assessment of responses from these detectors. The response of these detectors in active field applications has a certain error associated with it. The results (an alarm or no alarm) from these detectors have to be interpreted in the context of command actions that must be taken. The report discusses some examples of the risk assessment and illustrates how the information from the detectors must fit into the decision-making process.

The complexities of the protocols recommended in this report are primarily driven by two factors: the wide variety of backgrounds in which the CWAs are to be detected and the variable chemical nature of the CWAs that results in a range of infrared absorption bands on which their detection can be based. Detecting and measuring the spectral features of a CWA in the presence of thousands of background spectra using complex chemometric techniques requires statistical testing of simulants in both laboratory and field environments as well as laboratory testing of the various CWAs. It is these factors that result in the large amount of testing which has to be done to assure detector performance.

The committee's protocols do not call for field testing with CWAs, and the committee presents justification to show that this step is not necessary to provide statistically valid protocols.

SIMULANTS

Simulants are chemicals with lower toxicity than CWAs that can be used in the testing and evaluation of standoff detectors, in both the laboratory and the field. Two broad categories of simulants can be identified—spectral simulants and aerosol simulants.

For spectral simulants the spectra of these compounds should ideally have their maximum absorption within 20 cm^{-1} of the CWA of interest, and the spectra of the simulants should not have absorption bands stronger than the two strongest bands of the spectrum of interest.

Development of a generic simulant for CWA aerosols and droplets poses challenges that have yet to be solved. Rigorous simulation of a CWA aerosol or droplet requires matching (1) the optical properties of the material, (2) the physical properties that determine the manner in which the agent is dispersed, and (3) the dispersal method in the battlefield environment.

SUMMARY OF RECOMMENDATIONS

From its study of the problem, the committee makes a series of recommendations regarding appropriate test protocols and risk assessment in standoff detector testing and evaluation. Because they are inherently different, the two categories of detectors, passive and active, that are based on infrared spectral measurements require somewhat different test protocols.

TESTING PROTOCOL FOR PASSIVE DETECTORS

The committee concluded that the protocol recommended in the Battelle report is inadequate and that the assumptions on which it is based are incorrect. The major flaw in the protocol is the lack of information about the highly variable backgrounds that will be observed in the field. The committee proposes an alternative protocol that first acquires a large sampling of background spectra under a variety of field conditions. Then laboratory chamber spectra are taken of simulants, CWAs, concomitants, and possible interferents. Combining these, spectra can be synthesized that would be observed under field conditions with simulants or agents, concomitants, and interferents.[4] The large collection of synthesized spectra provides a database for the training and verification of a signal-processing model. This model can be tested first with known concentrations of simulants, interferents, and possible concomitants under ideal conditions in a chamber. If validated under these conditions, the same signal-processing model can then be tested in the field using simulants, interferents, and concomitants. Successful verification of this model using simulants in field measurements validates the transfer of the signal-processing model from the chamber to field and gives high confidence that a similar model for CWAs, if successful in chamber measurements, would transfer to the field without the necessity of field measurements using CWAs.

Recommendation: A detailed test protocol for testing and evaluating passive standoff detectors is recommended that recognizes the importance of the highly variable background in which such detectors will be employed. Simulants are used to provide validation of the signal-processing model in both the laboratory and the field. A similar model, developed with only laboratory testing of CWAs, can be used for the detection of live CWAs in the field.

Application environments for the detector are defined in the proposed test protocol.

---

[4]Concomitants are compounds that may be present under battlefield conditions—such as adhesives, thickeners, and propellants—whose spectra may interfere with detection of the target chemicals. Interferents are chemicals that in some way interfere with the detection of the target chemicals. This interference can be spectral, chemical, or physical.

TESTING PROTOCOL FOR ACTIVE DETECTORS

Active detection systems are inherently easier to model using physical data on the various components that make up the field of measurement in such systems. In fact, modeling is necessary to study the detectability of simulants and CWAs with active detectors. This results in a somewhat different approach to the testing of these devices.

Recommendation: This test protocol will be used to validate a model for the measurement and detection of CWAs in the field. The model will be based on measurements and understanding of background, simulants, interferents, and concomitants as well as characteristics of the instrumentation.

Validation of the model with simulants in the laboratory and field along with laboratory validation with CWAs will provide a high degree of confidence in the ability of the instrument to detect CWAs in the field under battlefield or homeland security conditions.

Limitation of application environments for these detectors will be predicted using the validated model. Detection sensitivity for the CWAs as a function of environmental factors in the model, such as levels of fog, dust, etc., can be predicted for the model and define the applicability limits of the detector.

RESEARCH AREAS TO SUPPORT TEST PROTOCOLS FOR STANDOFF DETECTORS

The committee believes that additional information and data are needed to fully support the proposed test protocols. These are described in Chapter 6. Among these research areas are the following related to the development of an algorithm for signal processing.

Recommendation: It is recommended that algorithm development be a multigroup effort that will result in robust, upgradeable, transferable software that will utilize multiple, differentiated chemometric approaches to the problem.

Robust algorithms are an essential element in the success of these detectors. This effort should ensure not only that the algorithm is effective but also that they can be modified as necessary by software experts in different locations as needed for the application.

DECISION MAKING AND RISK

A potential concern with the deployment of a standoff detector that has only been tested with simulants—but never with “live” CWAs— under realistic field conditions is that this might lower the degree of trust in its results. However, the principal detection challenge—whether for simulant or agent—is whether the spectral signatures can be unequivocally distinguished from background radiation and confounding spectral features associated with interferents and concomitants in the atmosphere. The rigorous test protocols developed by the committee provide the data necessary to develop algorithms that will be able to discriminate CWAs in the complex spectra acquired by a given instrument. If an instrument can achieve the demanding level of performance required by the applicable test protocol using simulants in both laboratory and field tests, there is a high degree of confidence that it would also detect CWAs in actual field conditions.

The ultimate proof of a standoff detector’s worth will be its performance under field conditions. The measure of the value of a detector for making command decisions in the field is to analyze the improvement in such decisions made with information from the detector. The detector’s value lies in providing information that will permit accurate prediction and decision to protect troops in an actual CWA attack, as well as the degree to which unproductive protective measures can be avoided when there is no attack.

Recommendation: Decision and risk training in the context of standoff detectors should be given to personnel who will be associated with the operation, use, and analysis of information generated by the standoff detectors in the field.

Good decisions do not consist of following the detector's alarms blindly, but in using the detector's information to improve the ability to characterize the chances of a CWA attack in a given threat situation. Understanding the false positive and false negative rates, the situations that can lead to them, and the pitfalls of deciding what to do based only on the detector's readings are all key to the effective use of the detector and to optimal avoidance of losses. Training in the proper use of detectors in the context of the larger decision-making setting would help make the best use of the detectors' information.

FIELD TESTING WITH CWAS

The value of open-air testing of detectors with CWAs, if any, would lie in the degree to which false positive and false negative rates are better characterized and understood than could be achieved by testing with simulants alone. Arguing against CWA testing is the fact that any such testing must necessarily be limited to a fairly narrow range of conditions and CWAs. Thus, testing with CWAs in the field would add little, if anything, to the confidence in the device. Such limited testing could also create an unwarranted degree of confidence if the detector "works" under field testing conditions that may turn out to be much more favorable than those actually encountered in battle.

Recommendation: Testing CWAs in the field is not recommended. The added value of information from such tests, recognizing that they would be limited in number, would not provide a significant improvement in the confidence level about instrument performance.

A detailed discussion of risk assessment and the value of information analysis is provided in Appendix C.

While the focus of this study was the testing and evaluation of standoff detectors for military applications, the protocols established and described may have applicability to the broader use of such detectors in areas such as homeland defense, for example. Such applications would require more knowledge of deployment strategies for detectors in such applications and tailoring of the recommended protocols for these different scenarios.

1

Introduction

The detection of chemical warfare agents is a topic that attracts a good deal of attention in the current environment of the war on terrorism. It is well, therefore, to point out at the outset what this report does and does not cover because the broad topic is a critical one with many aspects that need to be addressed urgently.

The Department of Defense (DoD) is responsible for ensuring that standoff CWA detectors will meet user needs under battlefield and homeland defense applications. In fulfilling this responsibility, the Defense Threat Reduction Agency (DTRA) of the DoD sponsored a study by the Battelle Memorial Institute (BMI), *The Use of Chemical Agent Simulants in Standoff Detection Testing*.[5] This study issued a final report dated October 2001 which recommended that "using simulants instead of CWAs provides an effective means for conducting outdoor operational testing of standoff detection instruments," Simulants are chemicals that are less toxic yet approximate the optical and physical characteristics of CWAs used in the detection scheme. Review of the report throughout the DoD establishment raised concerns around the validity of its conclusion based on the proposed test protocol and the supporting information provided in the report. The concern by the DoD that a testing protocol, absent field testing with live CWAs, might not adequately predict the response of these detectors to CWAs in battlefield and homeland defense applications seems intuitively but perhaps not scientifically valid.

DTRA requested that the National Research Council (NRC) undertake an independent assessment of appropriate protocols for testing and evaluating standoff detectors and their expected reliability. The Statement of Work drafted by DTRA that specifies the desired product of this study is given in Appendix A. Testing and evaluating standoff detectors for CWAs and assuring their reliability in actual field environments require the utilization of several technologies. To carry out this study, the NRC assembled

[5] A.R. Blackburn. K.S.K. Chinn, S.D. Fortney, W.A. Ivancic, A.K. Judd, B.D. Lerner, and J. Ontiveros. 2001. *The Use of Chemical Agent Simulants in Standoff Detection Testing*. Battelle Memorial Institute, Columbus, Ohio, CBIAC Contract No. SPO700-00-D-3180. CBIAC Task No. 75.

a group of leading experts in the fields of infrared spectroscopy, laser-based measurements such as light detection and ranging (lidar), chemometrics, neural networks, aerosols, atmospheric sciences, and risk assessment (see Appendix B). Of necessity, the study was carried out on a relatively tight schedule, with the first meeting of the committee on November 7-8, 2002, and a second on November 25-26, 2002. The committee had vigorous discussions and debates on the major issues associated with each of the testing protocols and came to unanimous agreement on the recommendations contained in this report.

This report is a narrow and very specific study of the testing and evaluation of infrared-type standoff detectors for CWAs in military situations.[6] It is not a broad study of various possible methods of detecting CWAs, nor is it a study of the use of such detectors in applications such as homeland security. Rather, it is a study of the best possible way to evaluate such detectors in a realistic way that will ensure that they will detect what they are supposed to, primarily in a wartime scenario. The specifics of field applications, while relevant in the deployment of such detectors, were important in the present study only in the sense that the study had to recognize the breadth of situations in which these devices would be used during and after the military deployment of CWAs.

Although this study is narrowly focused, the committee addressed a number of critical issues. The first issue concerned which types of detectors were to be included in any test protocols. The BMI report focused entirely on passive infrared detection since this technique has been used by the DoD for about 20 years. The committee determined that, while this was current technology, its logical extension was to consider infrared detectors based on lidar technology as well, since this technology was likely to be incorporated in such devices in the not-too-distant future. Hence test protocols were discussed by the committee in the context of these two types of instruments. There was no discussion of alternative methods of detecting CWAs using different technologies as this was considered beyond the charge to this committee.

The second critical issue was the value of field testing of these devices with CWAs. The BMI report tried to make a case for using simulants in field tests to avoid the use of CWAs in such tests. Unfortunately, the protocol recommended in the BMI report did not technically support this conclusion. The present study provides more complex but necessary test protocols that eliminate the need for field-testing with CWAs, a highly desired goal for the testing and evaluation of these detectors.

The third critical issue was the risk assessment of responses from such detectors. These detectors are designed to measure CWAs in the atmosphere. The response of these detectors, like the response of any analytical instrumentation (no matter how simple or sophisticated) has a certain error associated with it. The testing and evaluation of these detectors will provide an assessment of that error and how it is related to the various environments in which the detector is deployed. Nevertheless, the response of the detector, like the result of any analytical measurement, must be interpreted in the context of how that result is to be used. In the case of standoff chemical detectors, the results (an alarm or no alarm) have to be interpreted in the context of actions that should be taken by the troops in a military situation, or a population, in a civilian environment. Thus the report discusses some examples of the risk assessment and indicates in these examples how the information from the detectors must fit into the decision-making process.

The complexities of the protocols recommended in this report are primarily driven by two factors: the lack of predictable backgrounds in which the CWAs are to be detected and the variable chemical nature of the CWAs that results in a wide range of infrared absorption bands. First, the unpredictable

[6]The Statement of Work for the present study may be found in Appendix A.

nature of the backgrounds in which these detectors will be used and expected to give reliable alarm information requires that the test protocol is based on data from a very large number of varied backgrounds. The CWAs to be detected must be discriminated against (i.e., detected in the presence of) any type of background. With passive infrared detection, there is no way to simplify this step. Second, because of the wide range of chemical structures of all possible CWAs, the detection scheme must be based on the fact that each potential CWA has a unique infrared absorption spectrum. Thus the CWA must have one or more infrared absorption bands in the spectral window that are being measured that are unique and not found in the background spectra. Detecting and measuring these spectral features in the presence of all the background spectra using sophisticated chemometric techniques requires statistical testing of simulants in both laboratory and field environments as well as laboratory testing of the various CWAs along with complex signal processing. It is these factors that result in the large amount of testing needed to assure detector performance.

If testing and evaluation of this class of detectors for CWAs are to be addressed in a way that provides reliable detection equipment, certain additional information will be necessary. This report recommends research that will be necessary to support the recommended test protocols.

The committee's protocols do not call for field testing with CWAs, and the committee presents justification to show that this step is not necessary to provide statistically valid protocols.

2

Overview

Testing and evaluation of standoff chemical warfare agent (CWA)[7] detectors to certify their suitability and reliability for field applications present a number of challenges. Two major challenges are (1) these detectors cannot readily be tested in a field environment with CWAs without extensive precautions and legal approvals by top government officials, and (2) the wide variety of environments in which the detectors might be used may significantly affect instrument performance. Both issues must be addressed in any test protocol evaluating such detectors in order to certify their suitability for use in the field.

This report is confined to two types of standoff detectors, so-called *passive* and *active* detectors that are based primarily on the infrared spectral properties of the CWAs, although the protocol for active detectors could also be applied to other wavelength regions. *Passive* detectors are those that record the infrared spectrum emitted or absorbed by the CWA relative to the surrounding background with the only source of excitation energy being from the ambient background. *Active* detectors employ transmitted infrared laser radiation that is scattered back to a co-located receiver from aerosols in the atmosphere or from a topographic target. The laser energy is attenuated by the natural atmosphere, the CWA, and other aerosols and gases in the battlefield environment. In both cases the detected radiation is analyzed spectrally and temporally with sophisticated algorithms to extract the signal due to any CWA present in the field. However, there is a fundamental difference regarding the ability to predict the signal source term for the two approaches. For an active detector the signal can be mathematically predicted from the detector system characteristics and from the scattering and absorption cross sections of the aerosols and gases as well as the topographic target characteristics that are present, allowing a complete model to be developed of the measurement process and uncertainties. On the other hand, the source term for passive detection (radiation from a widely varying background) cannot be accurately predicted to develop such

[7]Appendix D contains a list of acronyms and a glossary of terms.

an end-to-end model. This fundamental difference between the analysis of active and passive detection requires significant differences between the two protocols for testing and evaluation of these two types of instruments. The two detector types will have inherently different capabilities. The passive detector will work best when the CWA is in the vapor phase, whereas the active detector has the potential of working better when the CWA is in the form of an aerosol.

The technology incorporated in the existing standoff detectors for CWAs and other devices under near-term development is based on the infrared spectroscopic properties of the CWAs and the differentiation of these properties against the background wherever the detectors are deployed. While such properties of CWAs are unique to the agent, the background in which such detection may be used can be extremely complex from a spectroscopic standpoint. Therefore, the infrared signal that is gathered by the detector must be extensively analyzed in real time utilizing chemometric and/or neural network techniques in order to differentiate a signal resulting from a CWA from the background and other interfering substances (either deliberately released to the environment or occurring accidentally). Adding to the complexity of the spectra is the physical nature of the test environment. CWA release can occur in several ways, including artillery shell bursts, release from aircraft, ground release, and other means. Each of these release mechanisms will likely result in a somewhat different physical form of the CWA and a different localized background signal. In addition, specifically designed mixtures of non-CWA materials could produce spectral properties similar to those of CWAs and might be used to "fool" (spoof) or even "blind" (saturate or completely attenuate the signal of) the detectors.[8]

While logically desirable, testing and evaluation of any test instrumentation under actual operational conditions is not always possible or practical. This is especially true for standoff detectors for CWAs. Using CWAs in field testing presents inherent technical, safety, and environmental concerns as well as significant political, financial, and time costs. The approval process for such testing under existing national laws and international treaties on chemical weapons is extensive. It has been estimated that the financial cost alone for a single field test with a CWA is approximately $136 million whereas a comparable test with a simulant is significantly lower, about $400,000.[9] Even under the best protocol, a single field test of a standoff detector with a single CWA will not provide sufficient information on the reliability of these test devices for the variety of conditions in which they are expected to operate.

Simulants, on the other hand, are relatively benign materials that can be tested in the open environment at considerably lower cost. Ideally, a test protocol could include field testing of a given detector with simulants coupled with laboratory crossover studies with simulants and CWAs in order to develop confidence in the detector for field detection of CWAs.

This report discusses and addresses the questions raised by the DoD with respect to the value of testing with simulants and CWAs in chamber and field conditions. The committee recommends test protocols for standoff devices and provides the supporting rationale for such protocols. The test protocols recommended in this report are intended to test standoff detectors for the broadest possible field applications. Failure to pass certain steps in the protocol with increasingly complex backgrounds, will define limitations of that particular instrument in its field of application. For example, the instrument may work acceptably in situations where the background is sky or sea, but not a background with buildings, trees etc. It may be the judgment of the government contractor that a particular instrument

[8]The committee considered both of these possibilities in its deliberations. The committee determined that these issues were more tactical than operational, and thus were outside of its mandate.

[9]Salvatore Bosco, DTRA, presentation to the committee, November 7, 2002.

while limited in its fields of application can still be useful for specialized deployment. Risk assessment of the proposed test protocols is provided within the context of the scientific test procedures being recommended but is also discussed more broadly in the framework of decisions that must be made in the field. Various tactical considerations of risk that relate to field strategies for the use of these detectors are beyond the scope of this study and are therefore not addressed here. Initially, this report addresses the test protocol and underlying assumptions recommended in the Battelle report. The recommended test protocols resulting from this committee's discussions and the rationale for them are presented subsequently. There are also identified areas of research and data compilation that are required to support the proposed test protocols. Standoff detectors that can pass the test protocols recommended by the committee will provide high reliability in field performance for the detection of CWAs. The challenge for instrument design and manufacture is to provide standoff detectors that will measure up to these protocols.

3

Recommended Test Protocol and Decision Tree for Passive Detectors

TEST STRATEGY AND ASSUMPTIONS

The use of simulants is the basis of the protocol described in the Battelle report for validation of the Joint Service Lightweight Standoff Chemical Agent Detector (JSLSCAD), the passive standoff detection system for chemical warfare agents (CWAs).[10] These are chemical species that have a similar spectral response to the CWA of interest, with a known factor relating the absorption cross-sections. Signal-processing models are developed in chamber measurements for simulants to establish a detection limit, both alone and in the presence of possible interferents. The detection limit for the CWA is then assumed to be related to the detection limits for the simulants by the factor derived from the relative absorptivities. The signal-processing model is then checked with simulants in live-field measurements. The assumption is that similar results would be obtained in live-field tests of a CWA once the factor of the relative absorptivities is taken into account.

The committee concluded that the protocol recommended in the Battelle report is inadequate and that the assumptions on which it is based are incorrect. The major flaw in the protocol is the lack of information about the highly variable backgrounds that will be observed in the field. The background for chamber tests is an extended blackbody source of known temperature, whereas the background for field measurements is likely to be highly variable and could include forests, hillsides, sky, grasslands, lakes, etc., for which the emissivity will not only be less than 1 but will almost certainly vary across the spectrum. If this high variability results in adjustments to the model for simulants developed from chamber testing where the backgrounds are constant (though adjustable in terms of temperature of the blackbody source), there is no way to adjust the model for agents without running live-field tests.

The committee proposes an alternate protocol that it believes will address some if not all of these

[10]For a brief description of the JSLSCAD and its capabilities, see "Joint Service Lightweight Standoff Chemical Agent Detector (JSLSCAD)." U.S. Army Soldier and Biological Chemical Command. Edgewood Area, Aberdeen Proving Ground, Maryland, 21010-5424.

concerns. The protocol involves the use of simulants but in a significantly different fashion. Since the highly variable background issue is likely to be dominant, the protocol starts there. The initial task is to obtain a large sampling of background spectra acquired under field conditions. Enough spectra must be taken to fully span the space of all background conditions likely to be encountered in use of the standoff detectors. Background spectra must include field interferents such as smoke, diesel soot, dust, and propellants. Laboratory spectra are then taken of all CWAs, simulants, concomitants (e.g., adhesives, thickeners, propellants), and possible interferents (e.g., smoke, diesel soot).[11]

The spectral data obtained in this way can be used with the acquired field background spectra to synthesize spectra that would be observed under field conditions with said simulants, concomitants, interferents, and agents present. The simulation of spectra over a range of temperature differentials will require knowledge of the approximate background temperature of the scene being used. This large collection of background spectra provides data for training and verification of a signal-processing model to extract concentrations from measured field spectra. Since the chamber has a constant blackbody background, this corresponds to an ideal representation of the field measurements. Thus with data from the chamber, the model can now be used to verify that various real mixtures of simulants, interferents, and possibly concomitants can be measured under ideal conditions. If the signal-processing model can be validated for a variety of simulants—representing the spectral properties of known or expected CWAs—from chamber measurements, it can then move on to field measurements of simulants with interferents present in the field using *the same signal-processing model.* Successful validation of the model using these simulants and interferents in field measurements validates the transfer of the signal model from chamber to field. This is the underlying basis for the use of simulants in this protocol. This result would give relatively high confidence that the signal-processing model for CWAs, if successful in chamber measurements, would transfer successfully to field measurements without actually making live-agent field measurements. Furthermore, field measurements with CWAs would not provide statistically significant verification of the robustness of the model unless a very large number of live-agent tests were conducted under a wide variety of conditions.

Infrared spectral emission and absorption are related linearly to both the product $C \times L$ (for low concentrations, i.e., the sample is considered to be "optically thin"), where C is the concentration and L is the path length, as well as the temperature differential, ΔT, of the CWA relative to the background. In field applications of such detectors, the CL product will provide important sensitivity limitations to the detection provided ΔT is not zero and ideally greater than 3°C.[12] The delivery and release mechanisms for CWAs (explosive shells, fogging from an aircraft, release from a compressed state) will likely provide sufficient temperature differential for second or minutes, to allow such detectors to detect the infrared emission and absorption of the released material.

PROPOSED TEST PROTOCOL

Flow charts of the test protocols are given in Figures 1 and 2 to make it easier to visualize the steps and decision points in the protocol. The following text and figures are keyed to be used together.

[11]Concomitants would also include hydrolysis products produced during the destruction of CWAs.

[12]In these applications, the sample is optically thin.

Vapor-Phase Measurements

Step:

1. First, a very large number of background spectra (on the order of thousands of independent spectra) should be measured under as wide a range of battlefield conditions as possible. Spectra might, for example, be collected during military training maneuvers in a variety of locales. While these background spectra are being measured, a videotape of the scene should also be recorded so that descriptive scene information can be used as a reference. No CWAs or simulants should contribute to the spectra. Backgrounds should have as wide a range of emissivity as possible (from near-blackbody to sky). Spectra measured with the JSLSCAD should be measured both at a resolution of 16 cm^{-1} while the detector gimbal is rotating and at a resolution of 4 cm^{-1} while the gimbal is static. If other instruments are developed in the future, corresponding measurements should be made at both low and high resolutions. This will permit a low-resolution rapid scan to sense the "presence or absence" of CWAs and a high resolution scan for final decision making. It is suggested that every time new nonchemical weapons are being tested, at least four JSLSCAD instruments should be mounted so that as many background spectra as possible can be acquired. The collected spectra should be divided into two sets—one set (A) for training the signal-processing model and one set (B) for the validation of the model, with Set A being about twice as large as Set B. The spectra in Set A should all have been measured from different sites than those in Set B.

2. Acquire reference transmission spectra of as many simulants, concomitants, and interferents as possible under lab conditions (i.e., using an incandescent source and with the sample contained in a long-path cell at a known partial pressure and temperature). Make up all samples to 101 kPa (760 torr) with dry nitrogen. These spectra should be acquired at a resolution no worse than 4 cm^{-1}. The original interferograms should be retained so that the spectra can be readily degraded to other resolutions without the need to re-measure the spectra. Ensure that simulants of low volatility are included in these reference spectra. In addition, spectra should be obtained at various temperatures to determine any temperature effects on key vibrational features.[13]

3. Using the data acquired from Steps 1 and 2, synthesize two sets of spectra of simulants (Set A and Set B) under field conditions so that a wide range of CL products and ΔTs are included, where C is the concentration of the sample (ppm), L is the effective path length (m), and ΔT is the difference in temperature between the sample and the background (°C). Some of the background spectra should have been measured when common interferents (e.g., smoke, diesel soot) were present. At least 50 times as many spectra should be synthesized in this way than there are background spectra in sets A and B, using at least 50 spectra calculated with different CL products and ΔTs for each background spectrum. Linear dependence of the data should be avoided. For half of the spectra in each set, one of the simulants should be present and for the other half no simulants should be present. Set A will be used for training and Set B for validation in a manner analogous to the protocol developed by Yang and Griffiths.[14] It is possible that a number of different sets with different ranges of CL products and ΔTs for each simulant will need to be synthesized in this step. In this case, single background types (e.g., sky) should be used first and complexity added at a later stage of the training.

---

[13] D. Qin and P.R. Griffiths. 1994. Minimization of Quantitative Errors for Analysis of Vapor-Phase Infrared Spectra Measured at Different Temperatures by Partial Least Squares Regression. J. Quant. Spectrosc. Radiat. Transfer 52(1):51-58.

[14] H. Yang and P.R. Griffiths. 1999. Application of multi-layer feed-forward neural networks to automated compound identification in low-resolution open-path FT-IR spectrometry. Anal. Chem. 71:751-761.

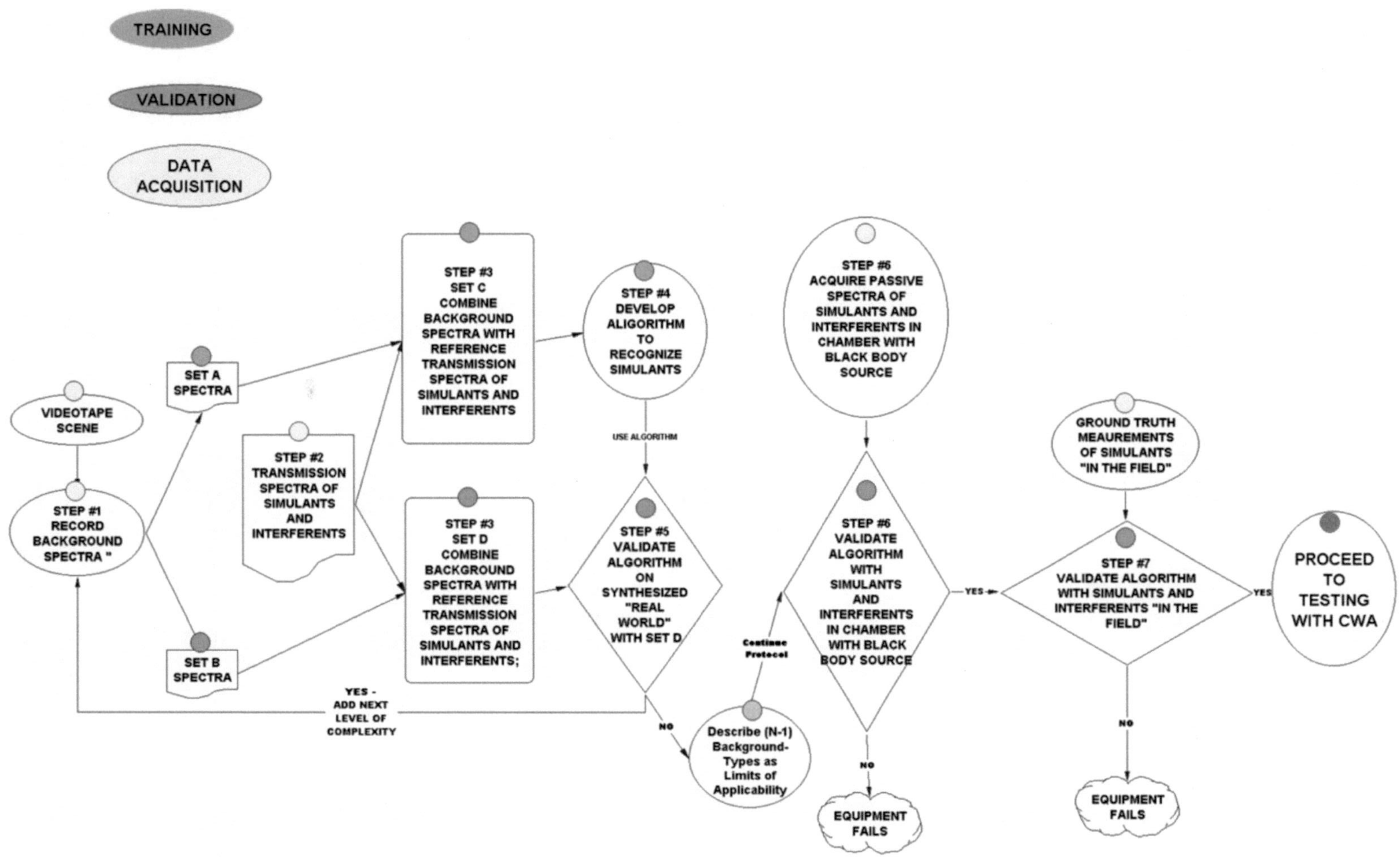

FIGURE 1 Test Protocol with Simulants for Passive Detectors; Pass 1 with vapor phase simulants; Pass 2 with aerosol.

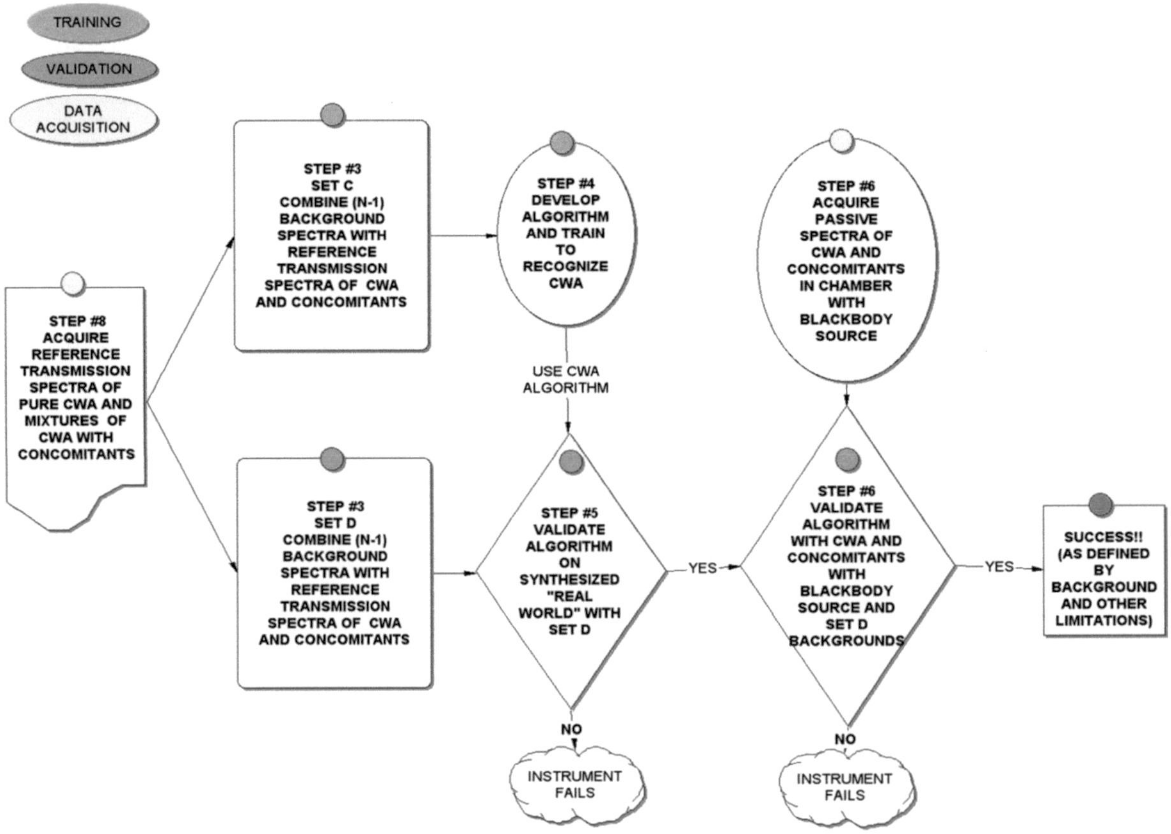

FIGURE 2 Test protocol with CWA for passive detectors.

4. Using the training spectra synthesized above (i.e., the calculated spectra in Set A), train an appropriate algorithm to recognize the presence of each simulant. The model(s) obtained in this way will be the basis of all subsequent tests. Initially, training and validation spectra should be calculated from similar backgrounds (e.g., sky and hillside backgrounds should not be included in the same set). The background spectra should then be expanded with the hope of calculating a universal model. If a universal signal-processing model is not feasible, software designed to classify background spectra into appropriate sets should be developed.

5. Validate the algorithms and models developed in Step 4 using spectra calculated from Set B. Find the limits of CL and ΔT for each simulant. If the algorithm is validated in this step, then repeat steps 1 through 5 using a more complex background set. This cycle is repeated until a background set is used that results in the failure of the algorithm. The last validation of the algorithm (n–1) thus defines the limits of applicability of the instrument in terms of its successful operation with increasingly complex backgrounds.

6. Acquire passive spectra of simulants and interferents in a chamber with a blackbody source to validate the model developed in Steps 4 and 5 using the same type of instrument used to collect the background spectra operated at the same resolution(s) used in the field.

7. Acquire spectra of simulants in the presence and absence of interferents (e.g., smoke, diesel soot)

in the field in order to validate the model developed in Step 4 and tested in Steps 5 and 6 using at least four different JSLSCAD instruments. At the same time, "ground truth" transmission spectra should be measured in two perpendicular directions through the simulant plume to find how CL varies throughout the series of measurements independently of ΔT. If possible, the temperature of the plume and the relevant background should be measured so that ΔT is known. (In practice, however, such measurements may be difficult to make.)

8. Acquire reference transmission spectra of pure agents and mixtures of agents along with expected concomitants (e.g., adhesives, thickeners, propellants), using the same conditions as in Step 2.

9. Repeat Steps 3 through 6, including agents of interest. The range of CL products and ΔTs should be specified by the Defense Threat Reduction Agency.

Aerosol Measurements

Repeat the tests established for vapor-phase measurements (Steps 1 through 9) for low-volatility simulants where aerosol formation is known to occur. A number of reference spectra of monodispersed aerosols with different particle sizes should be measured if possible. As wide a range of droplet diameters as possible should be covered. In addition, spectra of polydispersed aerosols with known ranges of diameters should be measured in order to compare spectra computed by adding the appropriate spectra of monodispersed aerosols of different sizes with the spectra of polydispersed aerosols. The simulants should ideally have physical and spectral characteristics similar to the CWA they are designed to simulate. If there is a need to choose between physical and spectral characteristics, it is more important to match the physical characteristics than the spectrum.

For chamber testing, make sure that simulants and concomitants with as wide a range of physical properties (Henry's law constants, viscosity, surface tension, etc.) are represented.

The "pass/fail" criteria for the performance of the instrument in this test protocol are the false positive and false negative rates, as defined in the Operational Requirements Document (ORD) for the device.[15] While this rate is defined in the present documentation, different acceptance requirements might be appropriate for different circumstances (i.e., military vs. civilian population protection, etc.). In all cases the acceptable false positive/false negative rates are the figures of merit for the test protocols.

Several points in the test protocol have decision nodes ("diamond" symbols in Figures 1 and 2). The first of these (Step 5) defines the scope of application of the instrument with each succeeding "Yes" loop. With each loop, additional and more complex background spectra are added to the "training" and "verification" sets. When or if the instrument finally fails to "pass" this node, the accumulated background scenarios for which the instrument performed acceptably define the application limits on the instrument. There is also an option at this node that a change to the ORD can be implemented before continuing the test to subsequent steps. This would either loosen or tighten the acceptance criteria, depending on the trade-offs desired in the utility of the equipment.

Failure to pass subsequent decision nodes in the test protocol indicates that the particular instrument is not capable of adequate field performance against the ORD. Successful performance through all subsequent decision nodes indicates that the instrument will provide field detection of CWAs with the expected false positive/false negative rates in the field conditions as defined in Step 5. As will be

---

[15]Operational Requirements Document (JORD) for Joint Service Lightweight Standoff Chemical Agent Detector (JSLSCAD), draft document.

subsequently discussed under the risk assessment (Chapter 7 and Appendix C), it needs to be recognized that the essentially infinite array of field conditions will likely cause some deterioration of instrument performance against the metrics used in testing, and there is no practical way of estimating this deterioration beyond the background testing that has been included in the signal-processing model, short of adding even more background spectra to the model.

4

Recommended Test Protocol and Decision Tree for Active Detectors (Lidar)

Active detectors in this context are called lidar, or light detection and ranging detectors. They utilize a laser excitation source that can be either fixed or tuned wavelength and a co-located receiver system to collect the scattered laser radiation (with or without wavelength shifting) from the atmosphere and/or topographic target. The successful design and use of this detection system, i.e., a system which sends out its own "interrogating" radiation from a source and records the reflected radiation that has been attenuated by interaction with the "field," depends on accurately modeling the system, including all components, both instrumental and in the field, as part of the model. The testing and evaluation experiments are then used to refine the model such that ultimately it can reliably predict field behavior of the instrument. References given later in this discussion provide examples of this methodology for other applications.

ASSUMPTIONS

This protocol assumes that:

1. field testing of chemical warfare agents (CWAs) can be avoided if adequate testing of simulants is performed in conjunction with a validated measurement uncertainty model that can realistically and reliably predict lidar performance for CWA measurements in battlefield or home defense environments;
2. no single simulant exists that can simultaneously emulate CWAs in optical and physical characteristics;
3. all battlefield conditions cannot be reproduced in field testing for evaluating lidar CWA measurement capabilities; and
4. steps in the protocol must be completed in sequence (unless otherwise stated), and if a step cannot be met to the degree required by the Operational Requirements Document (ORD), then a change to the ORD must be agreed to before proceeding to the next step.

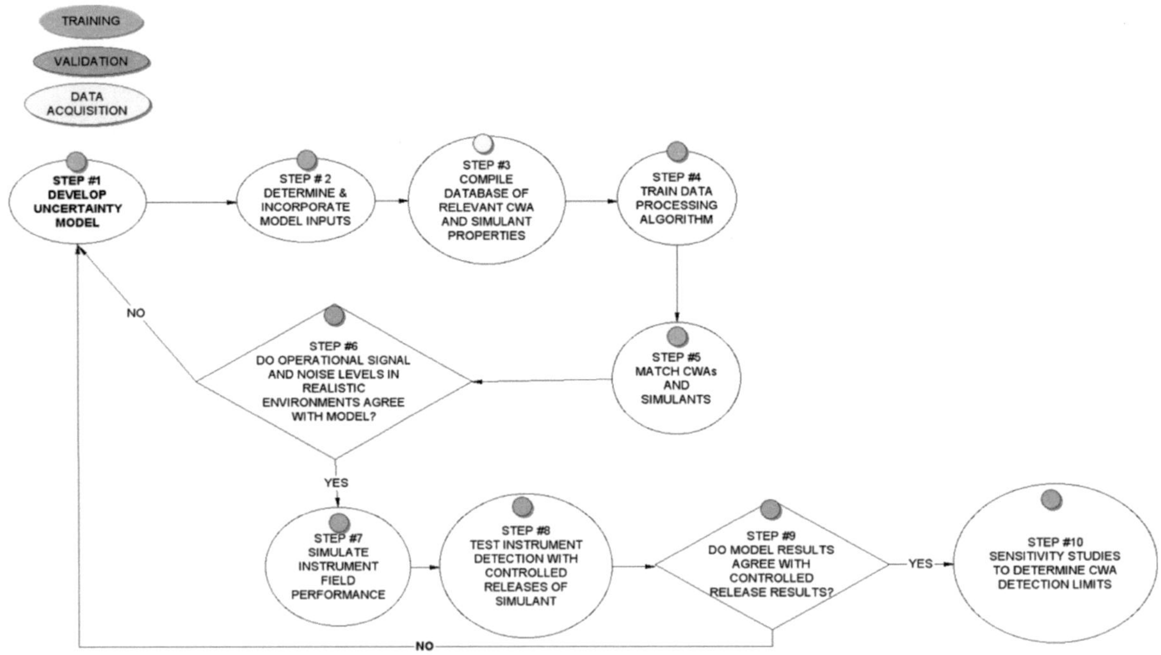

FIGURE 3 Test protocol for active detectors.

TESTING AND EVALUATION PROTOCOL (FIGURE 3)

Step:

1. Develop measurement uncertainty model. A comprehensive end-to-end measurement uncertainty model must be developed that includes all of the components necessary to realistically model the CWA measurement performance of the lidar system under conditions in which it is to be used. This must include *all* the factors that can contribute to uncertainties in the lidar CWA measurement performance. These factors entail battlefield environmental factors, including atmospheric scattering and attenuation from various sources; reflected and emitted background radiation sources; vapor, aerosol, and droplet characteristics of CWA releases and their spatial and temporal variation; spectral and spatial characteristics of topographic scattering; and lidar system characteristics as they pertain to measurement performance, such as transmitted energy per pulse, number of pulses averaged per measurement, wavelength stability of the laser source, range to measurement, size of receiver, field of view (FOV), collection and transmission efficiency (including optical filters), detector system noise, measurement sampling differences, and errors in application of the various lidar measurement techniques to be used. This model must be sufficiently comprehensive to be able to assess the sensitivity of the lidar system performance to variations in CWA release scenarios, changes in battlefield environments, meteorological and target variations, etc., *and* it must be able to be validated in comprehensive simulant experiments described below. Examples of this type of modeling for lidar measurements of gases were first discussed by

Schotland for water vapor measurements[16,17] and by Byer and Garbuny for measurements of pollutant gases in the infrared.[18]

2. Determine and incorporate model input parameters. All of the components of the model that are to be in place in the application of this lidar system must be accurately determined for the model to represent correctly the CWA measurement uncertainties of the system. These include the following:

- characteristics of all types of topographic scatterers that will be used by this system, including albedo characteristics, spectral dependence of scattering, and spatial variations (mixtures of surfaces within FOV);
- amount and wavelength dependence of background radiation from all sources within expected lidar FOV under conditions that the system will be used;
- relevant ambient environmental factors that would affect measurements such as atmospheric temperature, density, and wind;
- spatial distribution and physical and optical characteristics of all the different aerosols along the measurement FOV;
- spectral absorption characteristics of CWA, concomitant, simulant, and interferent vapors at the relevant spectral resolution for the lidar measurement;
- size distribution and refractive index of CWA, concomitant, simulant, and interferent aerosols and droplets;
- lidar system parameters that affect measurement of signal-to-noise ratio;
- technique-specific operational issues that could contribute to measurement error sources in the model, such as from spectral impurity, laser wavelength jitter, and Doppler broadening of backscattered signal, etc.; and
- modeling of the characteristics and uncertainties of the data analysis technique that will be used in the separation of multivariate components to arrive at the concentration of the CWAs such as principal component analyses and neural networks.

Note: Some of the atmospheric and topographic scattering and attenuation characteristics will be augmented by lidar measurements made as part of Step 6, and the CWA and simulant parameters will be defined as part of Step 3.

3. Compile a database of relevant CWA and simulant properties. Obtain accurate spectroscopic properties of CWA, simulant, and interferent vapors that are appropriate for the lidar system parameters (laser wavelengths and line widths, atmospheric conditions, etc.). Obtain accurate CWA, simulant, and interferent aerosol (and droplet) optical (complex refractive index) and physical (spatial and temporal variations in size distributions) characteristics in the range of atmospheric CWA releases to be considered as required by the ORD. (Note: The aerosol and droplet characteristics may be delivery mode dependent; if so, this information needs to be determined and tested in subsequent steps for each mode.) In most cases these measurements will require chamber tests using the lidar system to directly obtain the

---

[16]R.M. Schotland. 1966. Some observations of the vertical profile of water vapor by means of a laser optical radar. Fourth Symposium on Remote Sensing of Environment. Environmental Research Institute of Michigan. Ann Arbor, Michigan. Pp. 273-283.

[17]R.M. Schotland. 1974. Errors in the lidar measurement of atmospheric gases by differential absorption, J. Appl. Meteorol. 13:71-72.

[18]R.L. Byer and M. Garbuny. 1973. Pollutant Detection by Absorption Using Mie Scattering and Topographic Targets as Retroreflectors. Appl. Opt. 12:1496-1505.

optical parameters of the vapors, aerosols, and droplets, which are needed for model simulations of atmospheric measurements.

4. Develop a data-training algorithm. If a data-processing technique such as a neural network is used in the CWA measurement, it should be trained with the lidar in the high signal-to-noise environment of the chamber or through some other appropriate means (e.g., high-resolution spectroscopic technique).

5. Match CWAs and simulants. Based on the results of Step 3, determine which simulants are similar to which CWAs in their vapor absorption properties and aerosol and droplet scattering and absorption properties at operating wavelengths of the lidar and their aerosol and droplet physical characteristics.

6. Test system operational signal and noise levels in realistic environments. Operate the lidar system in a wide range of battlefield environments with ambient atmospheric scatterers, interferents, and a wide range of topographical targets over expected ranges. This will provide a test of the model to predict range-dependent signal and noise characteristics with different aerosol distributions, backgrounds, and topographic targets and their spectral dependencies. If the model does not accurately predict the lidar signals and noise, it must be fixed before going to the next step. It is expected that the atmospheric conditions and topographical characteristics will be sufficiently different than the assumed parameters used in the initial phase of the model development and that empirical changes to the model input parameters will need to be made to bring model predictions in line with actual conditions encountered in the battlefield environment. If the model cannot predict the uncertainty in the lidar measurements without a simulant present, return to Step 1 to determine and correct the fundamental reasons for the model deficiencies. Every step in the protocol must be examined to discover and correct the model or model input parameters.

7. Simulate instrument field performance. With the model, simulate an atmospheric measurement with a simulant release placed at the expected range for the lidar measurement in an expected battlefield condition (similar to what will be conducted in atmospheric tests). The absorption and scattering characteristics as well as the spatial and temporal variations of the simulant need to be included. The resulting modeled signals at the various lidar wavelengths are then input to the data-processing system for determination of the resulting measurement uncertainty for the simulant. A series of experimental simulations covering the range of simulants, mixture of simulants, and interferents need to be conducted to determine realistically the measurement uncertainty for a simulant. Experimental simulations should be conducted for all simulant testing to represent CWA vapor, aerosol, and droplet releases. Sensitivity studies should also be performed using the measurement simulation model to determine what are the most critical parameters to simulate in atmospheric tests to ensure the highest sensitivity to CWA characteristics.

8. Test instrument detection with controlled releases. Conduct atmospheric tests with simulant and interferent vapors, aerosols, and droplets used in chamber tests to assess lidar system performance at ranges of detection and concentration levels required by the ORD. The range of tests must cover the simulation of all CWA vapors, aerosols, and droplets required by the ORD under a wide range of atmospheric and background conditions with different natural and battlefield aerosol distributions, topographical targets, and CWA release modes, which might affect the encountered aerosol size distributions in different delivery modes. Measurement results will be compared to model predictions based on "ground-truth" observations simultaneously made during the tests by different sensor techniques for providing the atmospheric parameters, background conditions, simulant concentrations and distributions (vapors and aerosols/droplets), topographic albedo, etc., that are needed as inputs to the measurement simulation model and for direct comparison with the lidar measurement.

9. Compare model results with controlled release results. The results from Step 8 are compared with the measurement simulation model results. The model must be shown to reliably predict the performance of the lidar system for the atmospheric simulant measurements under a wide range of conditions. Once the model is determined to accurately predict the lidar performance, a series of sensitivity studies can be conducted using measured and assumed parameters for atmospheric conditions, battlefield interferents, backgrounds, topographic scatterers, ranges of measurements, CWA concentrations and delivery methods, etc., to predict the performance of the lidar system in satisfying the ORD. Preliminary studies of this type can be conducted with the model prior to this step; however, it is not until this step that it can be said with confidence that the model is complete enough or accurate enough to predict the lidar system performance for CWA measurements. If the model fails this requirement, return to Step 1 to determine and correct the fundamental reasons for the model deficiencies. Every step in the protocol must be examined to discover and correct the model or the model input parameters. Step 7 must be repeated to provide updated predictions of atmospheric simulation test results. If these new results indicate that the atmospheric tests should be done differently, Step 8 must be repeated. If no new tests are indicated, the results from the previous Step 8 can be used in comparison with the model results. If the model still cannot provide the required performance, a final repeat of Steps 1 through 9 can be attempted before terminating testing of the detector.

10. Conduct sensitivity studies to determine CWA measurement sensitivity. Once the model is determined to accurately predict lidar performance, a series of sensitivity studies can be conducted using measured and assumed parameters for atmospheric conditions, battlefield interferents, backgrounds, topographic scatterers, ranges of measurements, CWA concentrations and delivery methods, etc., to predict the performance of the lidar system in satisfying the ORD. Preliminary studies of this type can be conducted with the model prior to this step; however, it is not until this step that it can be said with confidence that the model is complete enough or accurate enough to predict the lidar system performance for CWA measurements.

Failure of an instrument to meet the desired false positive/false negative rates in Step 10 with the signal-processing model developed in Steps 1 through 6 is considered failure. However, the learning in Steps 7 and 8 with a particular instrument may suggest adjustments to the model that might improve it. If this is the case, the instrument may be retested with an improved model. If such improvements to the model are not possible, the instrument is rejected as not meeting the criteria.

POTENTIAL PROBLEM AREAS

1. The aerosol and droplet size distributions and their time histories for CWAs in various modes of delivery may not be known and cannot be established.

2. It may be difficult to generate representative size and number densities of aerosols or droplets of simulated CWAs in a large chamber for sensor testing.

3. Some aerosol simulants that have physical characteristics similar to CWAs may not have suitable refractive indices or spectral dependence to their absorption to simulate the CWA aerosols or droplets (see Chapter 5).

5

Simulant Characteristics and Specifics

As noted in the Introduction, the use of simulants in the testing and evaluation of standoff detectors is necessitated by cost and toxicity considerations that accompany the use of chemical warfare agents (CWAs). An ideal simulant is one that would mimic all the properties of a CWA except for its toxicity. This is never achieved. Thus, only the primary properties important to a particular test can be found in an appropriate simulant. In this context the next sections discuss *spectral simulants*; that is, chemicals that have the necessary properties to calibrate the infrared spectral detector, and *aerosol simulants*; that is, chemicals that mimic the physical dispersion characteristics of CWAs as necessary components of the background and delivery concomitants that the detector might "see."

SPECTRAL SIMULANTS

The type of compounds suitable as spectral simulants are those that give rise to similar *spectral* characteristics observed for a given CWA in the 10 μm atmospheric window, that is, between 700 and 1,300 cm^{-1}. This criterion does not necessarily mean that the *chemical* structure of the simulant should be similar to that of the CWA; however, in light of the fact that most functional groups give rise to absorption bands that absorb in a characteristic frequency range, this is quite likely. Ideally, the strongest band in the spectrum of the simulant in the 10 μm atmospheric window should have its maximum absorption within 20 cm^{-1} of the CWA of interest. If the signal model for the particular CWA places significant weight on one or more other wave number regions, the spectrum of the simulant should also have bands of similar strength within 20 cm^{-1} of these regions. The spectrum of the simulant should not contain other absorption bands that are stronger than the two strongest bands in the spectrum of the CWA. In terms of their physical properties, spectral simulants should be volatile enough that problems due to aerosol formation do not arise.

AEROSOL SIMULANTS

Simulants for vapor-phase detection must closely approximate the vapor pressure and spectroscopic absorption bands of the agent being simulated. Aerosols and droplets add considerable complexity to the task of simulation since the distribution of the liquid or solid particles with respect to size depends on material properties, such as viscosity, surface tension, and density, on the method of dispersal, and meteorological conditions. Aerosol particles affect remote sensing observations by contributing to the signal; that is, by infrared emission spectrometry as in passive infrared measurements or absorption in both passive infrared measurements and active measurements such as lidar, and DIAL (differential absorption lidar). Absorption and emission properties of aerosols depend on both composition and particle size relative to the wavelength being probed. For small or weakly absorbing particles the absorption efficiency is proportional to the mass concentration of the absorbing species. On the other hand, large particles will appear to be black, so the emission scales with their projected areas. The size corresponding to the transition between mass and area-dependent absorption/emission depends on the strength of the absorption; that is, on the imaginary part of the refractive index at the particular wavelength being observed. Quantitative analysis of line-of-sight optical absorption or emission measurements thus depends on the size-dependent absorption efficiency and the particle size distribution.

The influence of scattering must also be taken into account. For particles that are small compared to the wavelength of light, the scattering scales as d^6, where d is the particle diameter, so small particles scatter little light. For particles much larger than the wavelength, the scattering is proportional to the projected area of the particle. For particles with diameters comparable in size to the wavelength (i.e., those in the so-called Mie scattering regime), the scattering efficiency (the ratio of the effective scattering cross section of a particle to its projected area) is a complex function of particle size and material properties and can be substantially larger than unity. The contribution of aerosol particles to the total extinction coefficient of the atmosphere along the line of sight is, thus, an integral over the product of the size-dependent scattering efficiency and the particle size distribution function. Moreover, even nonabsorbing particles scatter radiant energy, so the background aerosol will contribute to and in many cases dominate this mechanism of degradation of the remote sensing signal. A further complication arises at aerosol concentrations that are sufficiently high that multiple scattering must be taken into account.

The presence of scattering particles is essential to active remote sensing methods such as lidar, becausethey scatter the probe pulse back to the detector. Return signals are generated by any atmospheric particles. Returns may be enhanced by the cloud produced when a chemical agent is dispersed, but other particles in the air will also contribute, including dust, smoke, fog, or clouds.

This preamble highlights the difficulty of developing a generic simulant for agent aerosols and droplets. A rigorous simulation would require matching (1) the optical properties (imaginary and real parts of the refractive index) of the material; (2) the physical properties that determine the way the agent is dispersed, including but not limited to viscosity, surface tension, density, and vapor pressure (and the temperature dependence of those properties); (3) the dispersal method; and (4) the battlefield environment.

Chamber experiments can be designed to evaluate the response of the sensor to the particulate agent but require special attention to the properties of suspended particles, particularly their tendency to be lost to chamber walls and to contaminate windows. The sedimentation velocity of a 10 μm droplet with a specific gravity of one is about 3 mm/s, and the sedimentation velocity scales as the d^2. Thus, a chamber must be of substantial height to enable extended measurements of the optical effects of suspended particles. Even then, particle losses may be substantial since any air motions in the chamber will bring particles close to the surfaces where they can deposit much more quickly than quiescent sedimentation may suggest. Alternatively, a continuous source of aerosolized agent might be employed, although continuous evaporation will complicate quantification of vapor-phase concentrations.

Issues in field validation experiments include (1) the ability to detect absorbing species that remain in aerosols or droplets, (2) the influence of particles produced by dispersal methods on the signals obtained from aerosolized agent, (3) the rate of evaporation of aerosols and droplets to produce vapors that may be detected, and (4) the influence of background aerosols on signals from both vapor and particulate agent. While exact duplication of the conditions of any battlefield release is unlikely, method validation is possible by "closure" studies in which the aerosol properties are characterized by in situ measurements of particle concentrations and size distribution, chemical composition, optical properties, etc., thereby enabling a direct comparison of modeled sensor response with field performance. Such closure studies have been successfully employed in many efforts to quantify the effect of aerosols (e.g., optically absorbing smoke, dust, and clouds) on the climate.[19,20]

---

[19]J.L. Ross, P.V. Hobbs, and B. Holben. 1998. Radiative characteristics of regional hazes dominated by smoke from biomass burning in Brazil: Closure tests and direct radiative forcing, J. Geophys. Res. Atmos. 103:31925-31941.

[20]D.R. Collins, H.H. Jonsson, J.H. Seinfeld, R.C. Flagan, S. Gasso, D.A. Hegg, P.B. Russell, B. Schmid, J. M. Livingston, E. Ostrom, K.J. Noone, L.M. Russell, and J.P. Putaud, 2000. In situ aerosol-size distributions and clear-column radiative closure during ACE-2. Tellus Ser. B-Chem. Phys. Meteorol. 52:498-525.

6

Research Areas to Support the Test Protocols for Standoff Detectors

The committee believes that there is a lack of information and data to fully support the proposed test protocols outlined earlier in this report. Accordingly, the committee identified the following specific areas of research that need to be addressed; results are needed for use in test protocols:

1. Measure and/or calculate aerosol and droplet spectra for background, simulants, and chemical warfare agents (CWAs) as input into the uncertainty model for active detector protocol. The aerosols in these studies should include both natural and anthropogenic sources such as combustion and evaporative emissions from human activities and accidental industrial releases.
2. Measure aerosol and droplet size distributions and densities resulting from various modes of delivery of chemicals. These should include simulants, CWAs, and concomitants. This will provide input for the uncertainty model for the active detector protocol.
3. Establish the optimum algorithm for processing data obtained by a passive infrared sensor such as Joint Service Lightweight Standoff Chemical Agent Detector (JSLSCAD). This may be instrument specific.
4. Establish the minimum number of calculation elements (algorithms, filters, networks, etc.) to process the spectral information resulting from the detector scans to ensure that "real-time" signal processing can be accomplished.
5. Develop a complete uncertainty model for the active (lidar) detector system. This capability will be useful to generate background types in the laboratory chamber.
6. Develop techniques to generate aerosols and droplets in appropriate laboratory chambers that are representative of the size and density distributions expected from various CWA delivery modalities.
7. Develop new simulants that resemble known CWAs in their spectral and physical properties but with low toxicity. Oligomeric variations of simulants may provide an avenue to permit systematic variations in such properties that can be evaluated for standoff detector evaluation and testing.

7

Decision Making and Risk

A potential concern with the deployment of a standoff detector that has only been tested with simulants under realistic field conditions—never with "live" chemical warfare agents (CWAs)—is that this might lower the degree of trust in its operation. A lack of confidence in the detector's performance, in turn, could reduce its value as a risk management tool for dealing with chemical threats to military operations. On the other hand, successful testing of a standoff detector with actual agents under a limited set of "open-air" or field conditions will not provide complete assurance that the detector would meet performance criteria under the myriad of weaponization and delivery modes as well as background conditions found on the battlefield. In fact, the committee is concerned that successful performance of a standoff detector with limited outdoor test sets with CWAs might lead to an unwarranted confidence in detector operation under variable field conditions and with various interferents.

Part of the difficulty regarding the decision-making process for detector testing and validation stems from perceived differences in the simulants and CWAs in terms of their inherent detectability in the atmosphere. Although simulants and agents have different physicochemical properties and toxicity, from a detection standpoint it is important that they have similar spectral signatures in the atmosphere that can be measured. The principal detection challenge—for simulant or for the agent—is whether those spectral features can be distinguished unequivocally from the background radiation in the ambient environment and from the confounding spectral features associated with chemical interferents in the atmosphere (e.g., naturally occurring compounds as well as those associated with battlefield environments). The committee has devised rigorous test protocols for testing and evaluation of both active and passive standoff CWA detectors. These protocols rely on a series of experiments and tests using simulants (in both chambers and open air) and live agents (only in chambers). These test protocols provide the data necessary to develop algorithms that will be able to detect CWA compounds in the complex spectra acquired by a given instrument in the field. Thus, the protocols explicitly address agent-specific spectra as well as the fundamental challenge encountered in instrumental detection of any airborne compound—the accurate identification of a given signal from background spectra. If a given standoff detector can achieve the demanding level of performance required by the applicable test protocol, there is a high degree of confidence that it would also detect CWAs in actual field conditions. Consequently, there is no compelling or overriding requirement for such tests. Live-agent testing under field conditions would provide minimal additional information for demonstrating detector effectiveness.

Another consideration affecting testing and validation is the anticipated role of the standoff detector in military operations. According to multiservice doctrine, risk management procedures should be utilized to guide the planning and execution of joint military operations.[21] Given this context, the operational role and performance of a standoff agent detector must be evaluated in terms of how it is used to manage risks that are detrimental to the successful completion of a given mission. The basic components of the risk management process include threat identification, threat/risk assessment, development and implementation of appropriate controls, and evaluation/revision of the risk management strategy.

Implementing this methodology for a mission that faces chemical warfare threats might include such details as atmospheric dispersion modeling of possible CWA attacks, prediction of potential casualties, and development of controls for risk management (e.g., modifying troop deployment, placement of detectors) to enhance the likelihood of mission success. In actual field implementation, additional information would come into play, such as measurement of meteorological conditions, intelligence on threats, and enemy location and activity. Moreover, detector performance can be addressed explicitly in the risk assessment process in terms of the likelihood (probability) that a detector would fail to detect the presence of a CWA (i.e., false negative) and the consequences of that failure to accomplishment of the mission. We note that standoff detectors are not required to be 100% reliable across all combat conditions and that, consequently, the level of trust placed in their operation always needs to be tempered with their potential limitations. Similarly, validation and field testing of detectors will never produce an infallible detector, but as a secondary control the military's risk management process should be sufficient to compensate for variable detector performance.

The ultimate proof of a standoff detector's worth will be its performance under field conditions, either on the battlefield or in antiterrorism applications. A commander in a threat situation must make decisions about the use of protective equipment, the way forces are deployed, and the actions and precautions that troops should take when faced with the possibility of a CWA attack. Protective measures are costly and hamper the fighting effectiveness of troops or the ability of a community to conduct its normal activities. In most settings a CWA attack is quite unlikely, and so precautionary measures impose costs that are unlikely to provide benefits. On the other hand, if an attack should occur, the costs to unprotected troops or a civilian population could be major, including severe casualties and risks to the mission. The value of a detector is in its ability to provide information about imminent exposures to CWAs, thereby allowing the commander to take avoidance or protective measures only when there is a high likelihood they are actually necessary. The way to measure the value of a detector is to analyze the improvement in the decisions made when it is able to provide information. The detector's value lies in the degree to which attacks on unprotected troops or civilian populations can be avoided as well as the degree to which unproductive protective measures can be avoided when there is no attack.

Any detector, no matter how well designed, will have some rate of failing to warn of actual attacks (false negatives) and some rate of sounding alarms when there is no attack (false positives). The false negative and false positive rates are key to determining the optimal decisions about adopting protective measures in a threat situation. The principal value of testing a detector beyond its design

[21] *Risk Management*. 2001. FM 3-100.12. Air Land Sea Application Center, Langley Air Force Base, VA.

phase is to establish the false positive and false negative rates. Variations in the setting, background, delivery mode, and CWA concentrations are key challenges to correct detector performance, and these will markedly influence the chances of false positives and false negatives. A valid testing protocol, therefore, must address the span of such conditions that are expected to be encountered by deployed units.

The value of open-air testing of detectors with CWAs, if any, would lie in the degree to which false positive and false negative rates are better characterized and understood than could be achieved by testing with simulants. Arguing for such live-agent testing is the fact that one is less sure of the operating characteristics of a detector toward real CWAs in the open air when testing is only on simulants than when testing is also on live CWAs. With simulants the characterization of detector performance is always an inference, never a direct demonstration, and the possibility exists that some unanticipated factor that differs between simulants and the CWAs might invalidate the inferences arising from simulant-only testing. This could result in apparent false reporting rates for the detector that are overly optimistic, most seriously if the detector would fail more often than estimated to sound alarms in the presence of real CW attacks in the field. Arguing against live-agent testing is the fact that any such testing must necessarily be limited to a fairly narrow range of conditions and CWAs, and so it may be harder to characterize the dependence of a detector on backgrounds, temperatures, humidities, and other factors in its ability to correctly sound alarms. If the detector "works" (i.e., sounds an alarm when live CWA is released) under fairly favorable conditions, this may provide a false sense of security in its abilities when conditions are less favorable or when they are significantly different from the setting in which the tests were carried out.

The appropriate way to judge these issues is to carry out a value-of-information analysis. This decision-analytic technique would aim at explicitly measuring the improvement in the field commander's decisions that result (and the consequent lowering of losses from inappropriate decisions) if additional open-air testing of the detector with live agent were to be done *versus* if it were not done.

A more detailed discussion of risk assessment and the value of information analysis, its application to the testing of standoff instrumentation, and issues that a decision maker needs to consider in using the field information from standoff detectors before taking action in the field is given in Appendix C. The committee concluded that risk management science has to be an integral part of utilizing the "hard" technical output from any monitoring instrumentation. One of the committee's recommendations underscores the importance of this issue.

8

Summary of Recommendations

After studying and discussing the problem and background, this committee makes the following recommendations:

1. A detailed protocol for testing and evaluating passive standoff detectors is recommended that recognizes the importance of the highly variable background in which such detectors will be employed. Simulants are used to provide validation of the transfer of the signal-processing model from chamber to field. This gives a high degree of confidence that a similar model for chemical warfare agents (CWAs), developed with only laboratory testing of CWAs, will transfer from chamber to field.

Application environments for the detector are defined in the proposed test protocol.

2. A different protocol is recommended for the testing and evaluation of active standoff detectors. This test protocol will be used to validate a model for the measurement and detection of CWAs in the field. The model will be based on measurements and understanding of background, simulants, interferents, and concomitants as well as characteristics of the instrumentation. Validation of the model with simulants in the laboratory and field along with laboratory validation with CWAs will provide a high degree of confidence in the ability of the instrument to detect CWAs in the field under battlefield or homeland security conditions.

Limitation of application environments for these detectors will be predicted using the validated model. Detection sensitivity for the CWAs as a function of environmental factors in the model, such as levels of fog, dust, etc., can be predicted for the model and define the applicability limits of the detector.

3. It is recommended that algorithm development be a multigroup effort that will result in robust, upgradeable, transferable software that will utilize multiple differentiated chemometric approaches to the problem. Robust algorithms are an essential element in the success of these detectors. This effort

should ensure not only that the algorithms are effective but also that they can be modified as necessary by software experts in different locations as needed for the application.

4. It is recommended that decision and risk training in the context of these standoff detectors be given to personnel who will be associated with the operation, use, and analysis of information generated by the standoff detectors in the field.

5. *Testing CWAs in the field is not recommended.* With the rigorous protocols given in this report, the added value of information from such tests, recognizing that they would be limited in number, would not provide a significant improvement in the confidence level about instrument performance.

While the focus of this study was the testing and evaluation of standoff detectors for military applications, the protocols established and described may have applicability to the broader use of such detectors in areas such as homeland defense, for example. Such applications would require more knowledge of deployment strategies for detectors in such applications and tailoring of the recommended protocols for these different scenarios.

Appendixes

Appendix A

Statement of Work

SUMMARY

The DoD is responsible for ensuring that standoff CWA detectors will meet user operational requirements in the real world. The DoD has previously tested and fielded only one such detector (the Remote Sensing Chemical Agent Alarm—M21 RSCAAL) and the DoD is currently testing a second (the Joint Service Lightweight Standoff Chemical Agent Detector—JSLSCAD). Concerns have been raised over plans to rely on very short range indoor "chamber testing" using real agents combined with long range field testing using chemicals chosen to simulate the real agents ("simulant testing"). If feasible, conducting open-air, live-agent testing would undoubtedly provide the highest level of confidence that standoff detectors will meet user requirements in the real world. However, doing so would require obtaining the appropriate political, safety, and environmental permissions at the congressional, executive, DoD, state, and local levels in accordance with 50 USC Sec 1512 and other statutory requirements. There are also significant costs and time requirements associated with live agent open air testing. On the other hand, relying only on chamber testing combined with field testing with simulants may cause the DoD to field a detector that will not be operationally effective. The criticality of this issue is not limited to war fighting. The DoD is planning to use these detectors for homeland defense of military installations, and there is potential that the nation will use these detectors for homeland defense of U.S. cities.

The Chemical and Biological Defense Program (CBDP) commissioned an examination of this issue by the Battelle Memorial Institute, which produced the report (to be provided), "The Use of Chemical Agent Simulants in Standoff Detection Testing," The Battelle report summarizes, "using simulants instead of chemical agent provides an effective means for conducting outdoor operational testing of standoff detection instruments such as the JSLSCAD system. When combined with live agent laboratory and chamber testing, it provides an integrated means of validating the JSLSCAD instrument, precluding the need for open air chemical agent testing." However, the deputy under secretary of the army for operations research (DUSA-OR), in the course of personally reviewing the Battelle report, concluded: "I believe the report's caveats about using simulants add up to great uncertainty regarding our ability to determine the operational effectiveness of standoff chemical detectors." The DUSA-OR recommended a review of the issue by the NAS, a body of experts recommended by the NAS, or a panel of experts convened by the Office of the Secretary of Defense. Other comments on the study will also be provided.

STATEMENT OF TASK

The NAS will answer the following:

1. What test protocols should be adopted to ensure that standoff chemical agent detectors will meet operational requirements and why. Consideration should be given to a variety of options to include chamber testing, chamber and simulant testing, and live-agent, open-air testing.
2. Identify the challenges associated with executing the recommended protocols.
3. Identify the risks associated with not doing open-air testing using live agents.
4. Using Multi-Service Tactics, Techniques, and Procedures for Risk Management as described in FM 3-100.12, *Air Land Sea Risk Management,* dated February 2001 (ref Annex D Risk Assessment Matrix), assess the risk associated with operationally employing standoff chemical agent detectors that have been tested at three possible levels: (1) baseline—live-agent chamber challenges combined with simulant open-air challenges; (2) baseline plus challenges in a test facility capable of enclosed long-range ,live-agent challenges; and (3) baseline plus live-agent, open-air challenges. If NAS does not feel qualified to assess the severity component of such a risk assessment, NAS may provide the probability component and defer risk assessment to the DoD.

SCOPE

This effort will use the JSLSCAD Derivation of Requirements document (Enclosure 3) to provide the operational context and baseline for standoff chemical agent detection requirements. This effort will consider both the uncertainties in meeting the requirement as a given and the uncertainties inherent in the way the requirement was derived.

Testing could well depend on the technology used for detection. This effort will consider the hardware and software technologies associated with current developmental detectors (M21 RSCAAL, JSLSCAD, Artemis, and MCAD). This effort will indicate which findings, if any, are specific to these technologies and which, if any, are inherently applicable to the problem of standoff chemical agent detection.

Appendix B

Committee Membership

Edwin P. Przybylowicz (NAE), chair, retired in 1991 after more than 35 years with the Eastman Kodak Company as senior vice president and director of research. He became assistant director of Kodak Research Laboratories in 1983 and was named director of research and elected senior vice president of the company in August 1985. He has served as a commissioner of the U.S.–Polish Joint Fund for Cooperation in Science and Engineering, a program that fosters the collaboration of Polish and U.S. scientists, chairing conferences and workshops on technology transfer in Poland, the Czech Republic, and Russia. From 1994 to 1996 he was director of the Center for Imaging Science at the Rochester Institute of Technology. Dr. Przybylowicz received a B.S. degree in chemistry from the University of Michigan and a Ph.D. degree in analytical chemistry from the Massachusetts Institute of Technology. He is a member of the National Academy of Engineering and has served on numerous National Research Council committees. He is currently an elected member of the International Union of Pure and Applied Chemistry (IUPAC) Bureau and its Executive Committee and past chair of the U.S. National Committee for IUPAC.

Edward V. Browell is head of the Lidar Applications Group in the Atmospheric Sciences Competency at the National Aeronautics and Space Administration's Langley Research Center (NASA LaRC) in Hampton, Virginia. He is an international authority on laser remote sensing systems and their application to atmospheric science investigations from ground-based, airborne, and space-based platforms. Dr. Browell received his Ph.D. in 1974 from the University of Florida, where he conducted both experimental and theoretical studies on the reflectance of laser beams from clouds and hazes. For the past 28 years, he has been at the NASA LaRC developing and applying lidar systems to a broad range of atmospheric investigations. His activities have primarily been focused on the development and application of airborne differential absorption lidar systems for studies of ozone, water vapor, aerosols, and clouds. He has participated in over 30 major airborne field experiments all over the world to study atmospheric processes in the troposphere and lower stratosphere and is a recognized international leader in this field. In 1991 Dr. Browell received the NASA Medal for Exceptional Scientific Achievement for his studies of fundamental atmospheric gas and aerosol processes, and in 1998 he was awarded the NASA Outstanding Leadership Medal for outstanding leadership in the development and application of laser remote sensing systems in the investigation of global atmospheric processes. Dr. Browell is author or coauthor of over 190 papers published in journals and books, and he is a fellow of the Optical Society of America and a member of the American Geophysical Union and the American Meteorological Society. Dr. Browell is also the chief editor for atmospheric technology for the *Journal of Atmospheric and Oceanic Technology*.

D. Bruce Chase is a senior research fellow with E. I. DuPont de Nemours and Company. He received a B.A. in mathematics and chemistry from Williams College in 1970 and a Ph.D. in physical chemistry from Princeton University in 1975. His research interests include industrial applications of vibrational spectroscopy, Fourier transform techniques and Fourier transform Raman spectroscopy, structure/orientation development in fibers, and near-field vibrational spectroscopy. Dr. Chase is a recipient of the Bunsen-Kirkhoff Prize, the Bomem-Michelson Award, the American Chemical Society Analytical Division Award in Spectrochemical Analysis, and the Anachem Award.

James A. de Haseth received his Ph.D. at the University of North Carolina, Chapel Hill, in 1977, before spending 18 months as a postdoctoral research associate at the University of Alabama. He joined the faculty of the University of Alabama in 1979. In 1983 he joined the faculty at the University of Georgia, where he is currently a professor of chemistry. Professor de Haseth's research is directed to the interface between chromatographic systems and Fourier transform infrared spectrometry. His research also encompasses the development and use of chemometric methods for data analysis. He has published in a wide variety of areas that pertain to Fourier transform infrared spectrometry.

Richard C. Flagan is a professor of chemical engineering at the California Institute of Technology, where his research involves investigations of chemical and physical processes in the atmosphere and the processing of advanced materials and devices. Dr. Flagan has received numerous awards for his work, including the Marian Smoluchowski Award from GAeF (1990), the David Sinclair Award from the American Association for Aerosol Research (1993), and the Thomas Bacon Award in Fluid-Particle Systems from the American Institute of Chemical Engineers (1997). He earned his bachelor's degree in mechanical engineering from the University of Michigan in 1969 and his S.M. and Ph.D. from the Massachusetts Institute of Technology in 1971 and 1973, respectively.

Peter R. Griffiths is a professor of analytical chemistry and chair of the Department of Chemistry at the University of Idaho. Dr. Griffiths obtained his bachelor's degree at Oxford University in 1964 and his Ph.D. also at Oxford in 1967. After short stints with two small businesses (Digilab and Sadtler Research), he held a faculty position at Ohio University from 1972 to 1982, before moving to the University of California, Riverside, where he spent seven years, prior to taking up his present position. Professor Griffiths's research is centered on the application of vibrational spectrometry to the solution of problems of analytical, environmental, and structural chemistry. Professor Griffiths's group is developing techniques that allow materials in hazardous or toxic waste sites to be identified remotely using optical fibers mounted on a robotic transport. Professor Griffiths has received several awards for his work, including the Fritz Pregl Medal of the Austrian Society for Analytical Chemistry in 1995; he was president of the Society for Applied Spectroscopy in 1994.

David W. Layton is division leader of the Health and Ecological Assessment Division at Lawrence Livermore National Laboratory (LLNL). During the early part of his 27-year career at LLNL, his research focused on the health and environmental impacts of geothermal energy development in the western United States. Later, he led a major assessment of the information available on conventional ordnance (i.e., explosives and propellants) in order to support studies of the impacts of weapons demilitarization. He was also one of the lead investigators of a DoD-funded project to revise the field water quality standards used by the U.S. military. After these projects, his research emphasis gradually shifted to the development of quantitative risk assessments of environmental contaminants. He has prepared risk assessments of hazardous gas releases, residual groundwater contamination at a Superfund site, plutonium-contaminated soils, heterocyclic amines in cooked foods, and nuclear wastes dumped in the Arctic Ocean. In addition, he has worked to improve exposure parameters for use in risk assessments. For example, he developed an innovative, metabolically based method for determining breathing rates and conducted research on the resuspension of particles indoors. More recently, he has initiated projects dealing with environmental assessments of transportation fuels and associated additives such as ethanol and methyl tertiary butyl ether.

Thomas A. Reichardt is a senior member of the technical staff at Sandia National Laboratories. He received his B.S. and M.S. in mechanical engineering from Purdue University in 1992 and 1994, respectively. He was awarded a Ph.D. in mechanical engineering from the University of Illinois at Urbana-Champaign in 1999. During the following year he was a postdoctoral appointee in the Diagnostics and Remote Sensing Department at Sandia National Laboratories. In 2000 he began work at his current position. He has conducted research on several laser-based measurement techniques for combustion diagnostics and environmental sensing, including laser-induced fluorescence, degenerate four-wave mixing, polarization spectroscopy, coherent anti-Stokes Raman spectroscopy, and backscatter absorption gas imaging. His current research interests include active and passive optical techniques for chemical species detection, tunable infrared laser sources, and development of a handheld device for detection of gas leaks.

Lorenz R. Rhomberg is an expert at Gradient Corporation in quantitative risk assessment, including pharmacokinetic modeling, and probabilistic methods, with special experience in chlorinated solvents and endocrine active agents. He is the author of several books and more than 50 articles on these topics. Before coming to Gradient, he was on the faculty of the Harvard School of Public Health, where he remains an adjunct professor, and at the U.S. Environmental Protection Agency (EPA). Dr. Rhomberg is active in professional groups and environmental policy development, focusing on current issues in the interpretation of toxicological data in human health risk assessment through service on panels sponsored by government, industry, and such organizations as the National Academy of Sciences and the International Life Sciences Institute. He is a member of EPA's Food Quality Protection Act Science Review Board and has participated in several recent Federal Insecticide, Fungicide, and Rodenticide Act Scientific Advisory Panel meetings concerning cumulative risk. Dr. Rhomberg earned his Ph.D. in population biology from the State University of New York at Stony Brook and his B.Sc. in biology from Queen's University in Ontario.

Appendix C

Risk Assessment in the Testing, Evaluation, and Use of Standoff Detectors

OVERVIEW OF RISK

A risk is an uncertainty about future gains or losses. That is, there are events in the future that may or may not occur, and different outcomes lead to different sets of benefits and decrements. The way to describe a risk is as a set of distinct potential future courses of events that together describe the classes of possible outcomes of interest. Each course of events can be characterized as a triplet, the elements of which are (1) an outcome, (2) the likelihood (i.e., the probability according to current judgment) that the outcome will occur, and (3) the consequences that follow if the outcome does indeed occur, expressed as the degree of gain or loss that will be suffered as a result.[22]

At least some risks can be reduced by taking appropriate actions to avoid or protect against potential adverse events. It may be possible to alter any of the three triplet elements; that is, (1) to block the threat (so that its appearance is no longer a possible course of events), (2) to alter the probabilities that different outcomes will occur (favoring outcomes with less expected loss), or (3) to mitigate the consequences of bad outcomes should they actually occur (so that losses will be less than they otherwise would be).

In many circumstances, however, it is not possible to avoid adverse possibilities completely. One is faced with an uncertainty about future losses because one does not know ahead of time what among the alternative courses of events will actually occur. Faced with an array of possible futures (each a triplet of event, likelihood, and consequences), one has an expectation of future loss that is the statistical *expected value* of this distribution of possibilities. The expected value is the average loss over all the possible events, each consequence weighted by its probability of actually occurring. Of course, only one set of events will actually happen, so the actual realized loss (or gain) after the fact will usually be less or more (perhaps enormously less or more) than the expected value before the fact, depending on whether the actual outcome was one with a big or a small consequence.

[22]S. Kaplan, and B.J. Garrick. 1981. On the quantitative definition of risk. *Risk Analysis.* 1(1):11-27.

In a threat environment where chemical warfare agents (CWAs) might plausibly be used in an attack, a commander must decide whether to implement certain protective measures for the troops. One such decision, for example, might be to have troops don protective clothing.[23] This decision must be made before the attack occurs. This is a classical problem of decision making under uncertainty. One must decide on an action (suiting up or not) *before* one finds out which course of events transpires (there is a CWA attack or there is not), and there are consequences to guessing wrongly. In making a risk management decision, the commander faced with an uncertain threat considers the possible futures, assesses their relative likelihood and the losses that will accrue under each eventuality, and makes a decision that is in some way optimal for the protection of forces and accomplishment of the mission.

If the commander decides to order the troops to don their protective gear (and if the protection afforded by the gear against the CWA is perfect), the losses consist of the reduced fighting efficiency of the unit and the decay of their physical well-being for future actions that wearing protective gear entails. This could be characterized as a moderate loss but one that is certain to follow from the decision to don protective gear. The impact on the troops exists whether or not there is a chemical attack. If the commander orders the troops to remain unprotected, however, the loss is zero if there is no CWA attack (since the troops are unencumbered by the suits), but the loss is very great if the attack does occur (consisting of the casualties caused by the agent and the threat to the mission's success).

The commander's decision is thus one of balancing the certain but moderate loss of suiting up against the expected loss from not suiting up. The loss from not suiting up is uncertain because it is contingent on an event that may or may not occur (the CWA attack).

Risk Decision Diagrams

The decision can be diagrammed as shown in Figure 4. The decision (shown by a square node) must be made before the uncertain event (shown by the circular node). In the threat situation the probability that a CWA attack will occur is estimated to be Pr(CW), and hence the probability that there will be no CWA attack is 1–Pr(CW). If the decision is to don protective gear, there is the loss of "performance decrement of suits" that is suffered whether or not a CWA attack ensues, but if no gear is donned, the losses depend on the outcome—no losses ("0") if it turns out there is no attack and "casualties and mission risk" if there is indeed an attack.

Clearly, the decision rests heavily on the perceived likelihood of a CWA attack, Pr(CW). If this likelihood is judged to be very small, a commander may decide not to don protective gear, risking the small chance of a big loss (if one guesses wrong) to avoid the certain chance of a moderate loss from suiting up. On the other hand, if a CWA attack is likely, there is a large chance of a large loss by not donning the gear, and it is almost certainly worth the performance decrement to avoid jeopardizing the troops and mission. This "decision analysis" formalizes the intuitive conclusion that, owing to the asymmetry of the consequences of guessing wrong about whether a CWA attack is coming, donning protective gear makes sense even when an attack is fairly unlikely, as long as it is credible, but that it does not make sense when the likelihood of an attack is remote. The question is one of how likely need the attack be and how different must the losses be under different scenarios.

[23]The range of tactical responses might cover minimal troop protective steps to having the entire force don complete protective wear. For purposes of illustrating the logic involved in risk assessment, donning protective clothing and taking no protective actions are the only decisions addressed in the present discussion.

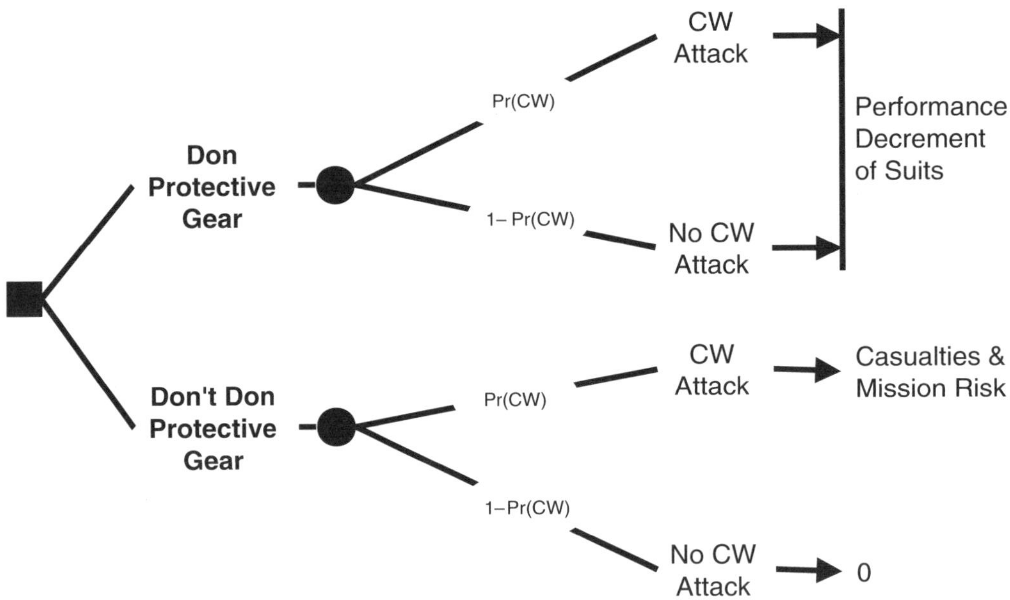

FIGURE 4 Command basic decision in the event of an alarm

In a business setting, when the same decision is made repeatedly and the gains and losses (in easily measured dollars) are accumulated over many similar episodes (e.g., underwriting insurance policies, speculating on stocks), the best decision is to take the choice that has the highest statistical expected value of gain (or the lowest expected value of loss). In the long run, this leads to the most success. In a military setting, however, it may be hard to express the different kinds of gains and losses in comparable units. Depending on the situation, the commander may wish to act in a way that is "risk averse" (i.e., accepting moderately increased losses in order to achieve a more certain albeit less favorable outcome) or in a way that is "risk taking" (accepting more chance of a significant loss in order to make possible a big gain if one guesses correctly). Nonetheless, the same sort of decision analysis framework applies.

Decision making under uncertainty can be improved if one can add information about the uncertain future event. For instance, a standoff detector can provide the commander with information about whether or not exposure of the troops to a CWA is imminent. If CWA exposure is imminent, a decision to don protective gear is clearly favored to avoid a highly likely large loss. If no CWA is detected (and if the commander can count on a useful warning of any attack), then going without protective gear is optimal. The essence of "perfect information" in a decision-analytic context is that it allows the decision maker to know which outcome of an uncertain process is going to occur *before* he or she has to make the decision. (This reverses the order with no such information, where the decision must be made before the outcome is learned.) Perfect information does not entirely obviate losses—if there is a CWA attack, the performance decrement to the suited-up troops from their gear cannot be avoided—but one can assure that the losses are the smallest possible for the outcome that actually happens.

The decision with perfect information is diagrammed in Figure 5. The standoff alarm either goes off or it does not. Since it provides perfect information, if it goes off, there is an imminent exposure to CWA coming, and so the decision to don protective gear is clearly favored. If there is no alarm, there is no imminent exposure, and the decision to go without protective gear is best.

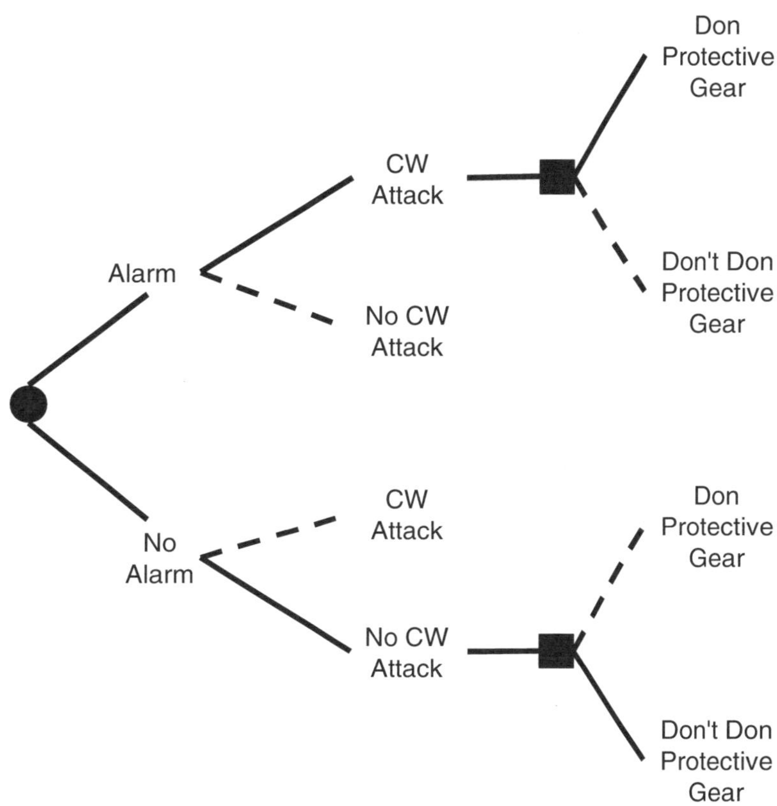

FIGURE 5 Decision tree with perfect information.

In practice, however, information is never perfect. First, it usually costs something to get. For standoff detectors this includes the costs of developing and testing the equipment, procuring the detectors, and transporting them, and devoting personnel to their maintenance and operation in the field. It also includes the "opportunity costs" of the other equipment that a unit must forgo in order to have the detectors, given limits on what can be transported and supported in the field. Whether these costs are worthwhile depends in part on the likelihood that the detectors will ever have anything to detect. Second, information is never perfect because it is rarely entirely predictive of future events. Typically, the information only *reduces uncertainty* about future events, but some uncertainty remains.

False Positives and False Negatives

For standoff detectors a detector in a threat environment will not always provide sufficient warning for every attack. There is some probability, for example, that a CWA shell will burst too close to the troops' position to provide time to don protective gear. There is also the possibility that the detector fails to detect the presence of CWA at concentrations that are nonetheless sufficient to inflict casualties. These are false negatives in that the detector's information that no CWA exposure is imminent proves to be wrong. Any decisions to remain unsuited made on the presumption that the detector's information is perfect will be in error if a false negative occurs.

A standoff detector may also have false positives, in which the detector warns of a CW attack that is not in fact occurring. This may be because of the "tuning" of the detector, but it could also occur if unanticipated environmental signals not investigated during testing cause alarms or if deliberate "spoofing" of the detector is carried out by the adversary, say by releasing chemicals that mimic CWAs, causing troops to suffer the performance decrement of operating in protective gear. Again, a decision made on the presumption that the detector is perfect will be in error, although the losses will be less than for false negatives. The losses will include the performance decrement attributable to protective gear and the gradual loss of confidence in the detector if false positives prove common (i.e., the "crying wolf" effect).

Weighing the Costs

When faced with a decision in the face of uncertainty, and when provided imperfect information, the decision maker still needs to trade off the costs of the various possibly wrong guesses. The information reduces but does not eliminate the possibility of guessing wrong. In the context of standoff detectors, a commander in a CWA threat situation must still decide whether to order troops to don protective gear. If the detector says there is no CWA present, the commander must decide whether this is a true negative or a false one.

In decision analysis this is approached as a problem in Bayesian probabilities.[24] There is a prior probability that the threat situation will entail a CW attack—this may be low for some settings and high for others, but it is the same assessment of the likelihood of attack that a commander would have to rely on if there were no detector. This probability is modified by the (imperfect) information of the detector, which either raises an alarm or does not. If the perceived probability of an attack is low and the detector fails to go off, the confidence that there is no CWA present is increased, and decisions to act accordingly are more likely to be correct than if the detector were not available. On the other hand, if the likelihood of an attack is perceived to be very low and the detector *does* go off, the likelihood that there is no real CWA attack drops precipitously, but it does not go to zero. There is still the possibility that this is a false positive alarm. If the plausibility of an attack is very low and the detector is known to produce false positives with some frequency, it may still be possible for the commander to decide that there is no attack and that the troops should not don protective gear. Similarly, if the perception that one is under CWA attack is very high and the detector does not go off, a commander might conclude that the failure to raise an alarm is a false negative and order the troops to suit up nonetheless.

Making the Decision

Risk analysts have an established methodology known as value-of-information analysis to analyze this kind of decision problem. The elements that enter this analysis are the description of risk in terms of the uncertain events, their alternative possible outcomes, the likelihood and loss consequences of each outcome, and the false positive and false negative rates of the source of information. Understanding the risk consequences of an imperfect detector (or one for which the degree of reliability is uncertain) is not simply a matter of looking at the consequences of a false positive or false negative alarm. The optimal decision is not simply to follow the detector's advice blindly, suiting up or not depending on whether it

[24]H. Raiffa. 1968. *Decision Analysis: Introductory Lectures on Choices Under Uncertainty*. Addison-Wesley, Reading, MA.

has raised an alarm or not. Instead, the optimal decision is the one that correctly balances the low likelihood of major losses from failing to don protective gear during an actual CWA attack against the highly likely but moderate losses from suiting up when no attack in fact occurs. How to make the decision to don protective gear depends not just on the detector's behavior but also on the prior likelihood of an attack, the detector's perceived false positive and negative rates, and the stakes for protection of the troops and the success of their mission.

The decision tree applicable to a value-of-information analysis is shown in Figure 6. In essence, the gear-donning decision diagrammed in Figure 4 can occur in two settings: (1) an alarm sounds or (2) no alarm sounds. What is different is that the probabilities that there will or will not be a CWA attack are changed—instead of the *prior* probabilities (based solely on the assessment of the threat situation) these are now the *posterior* probabilities of that assessment updated by the (imperfect) information provided by the detector. That is, if the alarm sounds, the likelihood that an attack is coming is Pr(CW|Alarm), the probability of an attack given that the alarm has sounded, and if the alarm does not sound, the likelihood that an attack comes nonetheless is Pr(CW|No Alarm).

Calculating these posterior or conditional probabilities can be done using Bayes's rule.[25] The calculation depends on the prior probabilities of attack as well as the false positive and false negative rates of the detector. For instance, if the false negative rate is designated α (and is equal to the probability of no alarm given that there truly is a CWA attack, i.e., Pr[No Alarm|CW]), and if the false positive rate is designated β (i.e., Pr[Alarm|No CW]), and if the prior probability of attack Pr(CW) is called q, then

$$\Pr(\text{CW}|\text{Alarm}) = q(1-\alpha)/[q(1-\alpha)+(1-q)\beta], \text{ and}$$
$$\Pr(\text{CW}|\text{No Alarm}) = (1-q)\beta/[q(1-\alpha)+(1-q)\beta].$$

Because of the added information provided by the detector, the commander's basis for guessing whether there will be a CWA attack is improved—without the detector this guess must be based on the assessment of the threat situation (expressed as Pr[CW]), but with the detector this likelihood either goes up considerably if there is an alarm (since Pr [CW|Alarm] is much bigger than Pr[CW]) or goes down considerably if there is no alarm (since Pr[CW|No Alarm] is smaller than Pr[CW]). As a result, the expected losses are reduced because it is much more likely that the commander can guess right about whether to order the use of protective gear. If there is an alarm, the decision to suit up has less chance of leading to the unproductive and unnecessary decrement of troops' effectiveness due to the cumbersome suits in the absence of a CWA to protect against; if there is no alarm, the decision not to suit up has less chance of leading to casualties and mission risk.

The difference between the commander's expected value of the loss in Figure 6 (with the detector) compared to Figure 4 (without the detector) is a measure of how much the commander can expect to gain by having the standoff detector. In the terminology of value-of-information analysis, it is the "expected value of testing information." (If it can be monetized, it gives the amount that the decision maker should be willing to pay to obtain the information.) That is, it is a measure of the degree to which use of the detector would be expected to lead to better outcomes (lower expected losses).

Whether the detector has a large or a small value of information depends on its false positive and false negative rates as well as on the probability that it will have something to detect; that is, the likelihood of a CWA attack in the threat situation of interest.

---

[25]H. Raffa. 1968. *Decision Analysis: Introductory Lectures on Choices Under Uncertainty*. Addison-Wesley, Reading, MA.

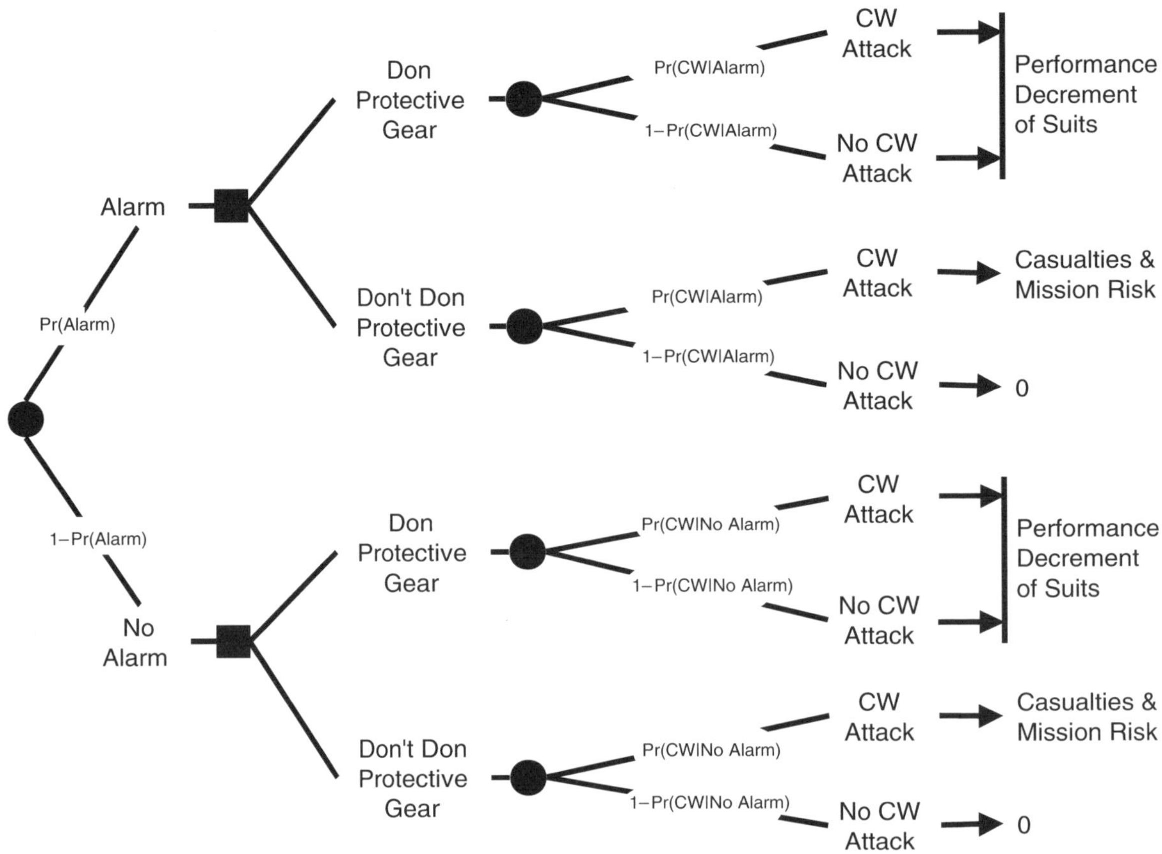

FIGURE 6 Decision tree in a value-of-information analysis.

Working through these questions was beyond the scope of the committee's charge, but we would urge that such an analysis be undertaken in the process of developing a risk management analysis of the use of any standoff detectors in the field and in the development of doctrine and training for detector-equipped units. It is an important part of understanding the risk management consequences of a standoff detector's reliability, and hence the risks of field testing the detector with live CWA or relying only on stimulant testing. Such a value-of-information analysis is the appropriate approach to answering the committee's Task 4—to assess the risks associated with operationally employing standoff CWA detectors that have been tested at different levels (see Appendix A—Statement of Work). The risk consequences of the detector's performance will be realized in its use in the field, and those consequences will depend on how a commander makes decisions based on the detector's outputs.

The Value of Information

It is clear from the above discussion that the key element about detector design and testing is the characterization of false positive and false negative rates. These rates are the quantitative characterization of the reliability of a detector, and alternative testing protocols will differ in the degree to which the uncertainty about a detector's field performance can be reduced. The way to approach Task 4 is to evaluate the differences in value of information provided by the detector, which in turn depends on the false positive and false negative rates that emerge from the testing. Indeed, the purpose of the testing protocol should be to characterize the false positive and false negative rates of the detector because they would affect battlefield decisions. This point bears some elaboration.

First, one must distinguish between the points of view of detector developers and field commanders. For the developers, false positive and false negative rates have to do with measuring the ability of a detector to meet technical specifications about detecting concentrations of a fixed set of CWAs in a defined set of testing conditions. From the point of view of a field commander, however, false positive and false negative refer to the correspondence of the detector's status to the actual threats to the troops. If an adversary uses a novel CWA that the detector was not designed to see or if a shell lands amid a concentration of troops and exposes them all just as the detector goes off (i.e., with no prior warning), these are "false negatives"—failure of the detector to provide sufficient warning—from the point of view of the field commander's decision analysis regarding use of protective gear.

Second, a testing program cannot test every possible environment and setting. If the wider variety of settings in the field lead to more frequent failures to perform accurately, this is a case where the actual operational false positive and false negative rates may be higher than those measured during testing. A good testing program will try to minimize this problem by testing over a wide variety of environments.

Third, the risk consequences of false negatives and false positives depend on the field commander's decisions and how they are influenced by the detector, and those decisions will be based on the *perceived* false positive and false negative rates, which may differ from the actual rates. Less than optimal decisions arise both when the decision maker overrates the detector's reliability and when he or she underrates it.

Fourth, there is typically a design trade-off between false positive and false negative reports. Although one strives to minimize both, in practice there is a limit to the degree that this can be accomplished. Since failing to warn of an actual CWA attack is a much more serious error than falsely warning of an attack, the detector should be "tuned" so as to minimize false negatives even at the expense of increasing the false positive rate. The optimal ratio of false negatives to false positives depends on the relative consequences of the two types of errors or, more precisely, the relative consequences of the erroneous decisions that are made in view of the detector errors (and as noted above, the rate of decision errors depends on more than just the detector's performance). The projected risk management consequences of the detector's field performance thus become an important part of the design process, influencing the optimal tuning between false positives and false negatives.

Comments on False Positive Rates

It is important to appreciate that, when true positives are rare, even for a very small false positive rate, most of the positives are false ones. In the present context, if actual CWA attacks are rare (as they have been historically), then most detector alarms in the field will be false positive. Specifications for detector design that call for a certain small proportion of alarms to be false are not useful, since the proportion of false alarms is mostly controlled by the frequency of true CWA attacks, something that is beyond the control of detector designers. In any case, a favorable proportion of true positives is only achievable by increasing the rate of true CWA attacks, which is not a good policy goal. The only other way to make a small fraction of alarms be false alarms is to design the detector to have an extraordinarily small false positive rate, but this must be achieved at the cost of increasing the false negative rate. Since the consequences of a false negative are much greater than those of a false positive, this is a bad policy as well. To illustrate, if the true frequency of CWA attacks on units with detectors is 0.1% (i.e., 1 in

TABLE 1 Calculations of Alarm Incidents Under Two Different Scenarios

True Frequency	True Incidents	False Positive Rate	False Negative Rate	No. of Alarms	True Positive Alarms	False Positive Alarms	False Negatives	True Positive
1/1,000	100	1%	1%	1,099	99	1,000	1	9.0%
1/1,000	100	0.1%	10%	190	90	100	10	47%

True frequency = frequency of CWA release in threat situations
True incidents = number of threat situations times true frequency
False positive rate = percentage of analysis resulting in alarm
False negative rate = percentage of actual incidents not detected
No. of Alarms = number if signals given by the instrument that a CWA is present
True positive alarms = number of true incidents alarmed by instrument
False positive alarms = number of alarms that do not coincide with incidents
False negatives = number of incidents that did not trigger alarm

1,000 threat situations lead to an actual attack) and the false positive and false negative rates are both 1%, then 1.1% of threat situations will have a positive alarm, and only 9.0% of these alarms will be true positives, while 90.9% will be false positives. If in response to this result the detector were retuned to achieve a 10-fold lower false positive rate (0.1%) at the cost of a proportional rise in the false negative rate (to 10%), the fraction of alarms that are false will be only 53%, but 10% of actual CWA attacks will not set off the alarm. Table 1 gives an illustration of how these changes in the false positive/false negative rate would give different decision criteria for the commander in the field. For illustrative purposes assume that there are 100,000 threat situations.

The above discussion concerns the risks stemming from the performance of the detector, and it emphasizes the role of testing in determining false positive and false negative rates. One can also speak of the risks associated with different modes of testing the detector. Here the risks arise from poorly estimating the false positive and false negative rates for CWA detection.

Value of Field Testing with CWAs

The testing protocol can also be looked at as a value-of-information question. We considered above the value of the information provided by the detector to the field commander, who has to make a choice in the face of imperfect information about the presence of CWAs. Now we can consider the value of information provided by field testing the detector with live agent and not just simulants. We have the choice to make whether to deploy the detector after simulant testing and chamber studies or to pay the higher cost, as outlined in the body of the report, if field testing with live agent is also carried out. The uncertain future event is how well the detector will perform in battlefield use. In particular, the uncertainty is how well the testing program has established the actual false positive and false negative rates. The value-of-information question is how much the additional field testing with live agent improves our confidence in the false positive and false negative rate characterization of the testing program.

A decision tree diagram could be made in which each branch represents an alternative testing protocol (e.g., with and without final open-air, live-agent testing). From each such branch tip would extend a sub-tree corresponding to the whole tree of Figure 6; that is, each subtree showing the decision-analytic problem that follows after a detector is deployed with a different set of testing procedures. The subtrees will differ only in their false positive and false negative rates, but these differences will affect the calculation of the expected losses a commander would have if he or she had to depend on a detector with the given values of α and β.

It should be noted that, once a detector is deployed, actual field experience would supplant the testing as the best basis for judging the detector's value. Information on the false positive rate in the field will accrue more quickly since most settings do not have CWAs present. Given the rarity of actual CWA attacks, the false negative rate will be difficult to determine through experience, and no additional information on this rate (beyond that provided by testing) will accrue until actual attacks occur. The losses from a false negative (casualties owing to failure of the detector to warn of an actual attack in the field) may be high, but they will not be suffered repeatedly. If it ever becomes clear through experience that the detector has a substantial false negative rate, commanders will cease to rely on it. On the other hand, the losses from false positives (suiting up when it is unnecessary) will likely be suffered repeatedly since the fraction of alarms that are true is necessarily small (as shown above) and since the costs of failing to heed a true alarm are large.

Identifying the Risks

The risks associated with testing or not testing with live agent are triplets of something that could go wrong, the likelihood that it will go wrong, and the magnitude of the consequences. The value of information provided by live-agent testing is judged by the degree to which the probabilities of the bad outcomes are reduced by the additional testing, and the resulting decrease in the expected value of losses. Briefly, the risks are:

- The detector will not work at all with real CWAs on the battlefield. This seems very unlikely if the testing program with simulants is well done, but if it were so it would have very large consequences since the detector would be falsely assumed to be showing the lack of CWAs when there may indeed be some. The actual risk depends on the likelihood that there is a CWA attack. If detectors that cannot actually detect the agent are deployed and their negative readings believed, this is only a problem if there is a true CWA attack. If there is no attack, there is no consequence. Thus, the ultimate risks of poor detector performance are all contingent on their ever being put to the test in a real battlefield situation, and the estimation of consequences of detector failure need to include the likelihood of a real challenge.
- The detector has a higher actual false negative rate than the testing has led us to believe. This will cause commanders to place too much trust in the lack of alarms. This seems to be a fairly likely outcome—the challenge of detecting agents in the variety of field conditions that may be encountered may not be adequately probed by the testing.
- The detector has a lower actual false negative rate than the testing has led us to believe. This seems unlikely. The consequences are likely to be good—better than expected detection and (if the alarms are believed even though they are thought to be less reliable than they indeed are) more likelihood of a decision that avoids CWA casualties. There may be more tendency to suit up in the absence of an alarm, however, since the lack of alarm is being given insufficient credence.
- The detector has a higher false positive rate than the testing has led us to believe. This seems fairly likely. The immediate consequences would be that units often unnecessarily don protective gear and suffer the performance decrements that accompany this. In the longer run, as it becomes clear from experience that most of the alarms are false, there may be a tendency to undervalue and ignore the alarm information (the "crying wolf" effect), leading to vulnerability to a real attack.
- The detector has a lower false positive rate than testing has led us to believe. As with a lower false

negative rate, this seems unlikely. It would lead to less than optimal readiness to don gear since the alarms would be given insufficient credence, and true attacks may find troops having ignored an alarm because it was perceived as likely to be a false alarm.

Beyond the question of the operational risks, one can examine the risks associated with the testing procedures themselves. These are adverse events that could happen during the testing. They can be broadly divided into risks not associated with detector performance and those that are associated with performance.

Risks of testing not associated with detector performance include the following:

- *Environmental risks.* These include human and ecological health risks associated with releasing chemicals into the environment. The human health risks would probably be assessed as the consequence of accidents or unintended and unexpected turns of events during a chemical release. The analysis would have to include identifying the potential failures and their likelihood, as well as the transport of chemical away from safe areas and the likelihood of human exposures of various degrees. The ecological risks may also be a result of unintended turns of events, but there may also be some fairly certain impact in the test area itself. There will be environmental risks for testing live CWAs in the open air, but there will also be risks from the additional testing of simulants that would be necessary if live-agent testing were not carried out. Although the simulants are markedly less toxic, they are not all entirely benign, and their potential environmental toxicity should not be overlooked.
- *Political and public relations risks.* These are risks of unfavorable public or political reaction to the testing program, with consequences to the image of the program and possibly to its funding and mode of operation. Again, there are risks associated with both live-agent testing and lack of such testing. Some constituencies will be more concerned about releases of live CWAs during tests, while others may be more concerned about the military preparedness consequences of failing to do live-agent tests.
- *Psychological and risk management risks.* If live CWA testing is not conducted, there may be lack of confidence in the detector's reliability among users, and this will affect the decisions they make in the field when detectors are deployed. Live-agent testing would probably increase confidence in the detectors on the part of troops. It may even result in overconfidence in the detectors, if the false positive and (especially) false negative rates are overly optimistic. The operational risks of the detectors (and hence their military effectiveness) will depend on the perceived reliability and the soundness of decisions made in view of the detector's performance. Suboptimal decisions are made if users are either overconfident or underconfident in the detector's abilities, and live-agent testing will affect this degree of confidence.

There are also risks associated with the technical performance of the detector. If live-agent testing is not conducted, these include the specific possibilities for failure of the logic by which simulant-only testing is presumed to provide sufficient understanding of the detector's performance. For instance, if actual CWAs have chemical interactions with ambient substances when released into the air, if this property is not suspected, and if simulants lack this property, the validity of the simulant testing may fail—a failure that would only be uncovered when detectors do not perform as expected in the field, with consequent impact on troops. It is worthwhile to attempt to list the factors that could compromise the validity of simulant testing, but that would be revealed by live-agent testing and to estimate the likelihood that each problem might actually arise together with the consequences of failing to appreciate it. This list should include not just likely problems but also problems deemed unlikely yet possible. Some such factors include:

• unexpected chemical interactions of CWAs with ambient substances,
• unexpected properties of CWAs when released from exploding shells,
• interferents whose spectral properties match those of a CWA in the spectral region being monitored, and
• the role of aerosols in detectability of agents and differences in vapor pressure of simulants and live agents.

It is also worthwhile to consider risks that arise when live-agent, open-air testing is done, but the validity of such testing for the needed purposes is compromised. The point of a live test would be to test both the validity of the simulation and the performance of the instrument. This is why a few tests cannot make a statistically significant contribution. An alarm from a clear and unambiguous live agent release (a "successful test," in that the detector "worked") is not very informative about the actual performance to be anticipated in real use. Such a test can only uncover gross flaws in the detector—if it fails an easy test it is bound to fail in harder situations. Here the problem is with overinterpretation or overgeneralization of the testing to conclude that the properties of the detector are well characterized when in fact they are not. If the detector is not challenged during testing with sufficiently difficult detection tasks, or if conditions that would apply in the field (and that make the detector tend to fail) are not appreciated and therefore not tested during the live-agent tests, false confidence is engendered from the program of live-agent testing.

The comparability of the simulants and the live agent in the testing situation is likely to be contingent on the setting, background, climate, delivery mode, and so on. If correspondence of the detector's performance with simulants and with live agents is to be considered telling about the detector's abilities, then the correspondence ought to be tested over a range of settings.

SUMMARY

In sum, the purpose of testing is to establish the false positive and false negative rates of the detector in a way that applies to the performance of the apparatus in actual deployed situations. The value of the detector depends on these false positive and false negative rates, and the soundness of decisions based on the detector's readings depends on accurate appraisal of the false positive and false negative rates. The risks associated with testing per se are the possibilities of unrecognized or uncharacterized factors that could compromise the validity of the measurement of false positive and false negative rates. The test protocols rigorously followed are intended to provide good estimates of the false positive/false negative rates for CWA detection in the field based on the protocol-specific signal-processing models.

Appendix D

Acronyms and Glossary of Terms

Active Detectors: detectors that create an optical signal by probing a scene with a light source (i.e., a laser); lidar is an example of such a detector.
Aerosol: small solid particles or droplets suspended in a gas or vapor.
Albedo: the ratio of light scattering to the sum of absorption plus scattering.
Artemis: a CWA standoff detection system designed to augment existing CWA detection systems to provide a theater-wide capability of near real-time detection and warning.

Blackbody: an object that completely absorbs incident radiation of all wavelengths and emits radiation according to Planck's law.
BMI: Battelle Memorial Institute.

CBDP: Chemical and Biological Defense Program.
Chemometrics: the application of any of a variety of multivariate algorithms to data from chemical instruments to perform qualitative or quantitative analysis.
Concomitants: compounds that may be present under battlefield conditions—such as adhesives, thickeners, and propellants—whose spectra may interfere with detection of target chemicals.
CWA: chemical warfare agent.

DIAL: differential absorption lidar—lidar technique based on transmitting at least two wavelengths of light that are absorbed to different degrees by the target gas.
DoD: Department of Defense.
DTRA: Defense Threat Reduction Agency—a department within the DoD.
DUSA-OR: deputy under secretary of the army for operations research

FOV: Field of view—angular breadth that can be imaged by the system (radians).

Ground truth: measurement that allows a particular parameter to be accurately known.

Henry's law: the pressure of a gas above a solution is proportional to the concentration of the gas in solution.

Interferents: chemicals that in some way interfere with the detection of target chemicals; this interference can be spectral (spectral signals that are similar to the target chemical), chemical (materials that react in the atmosphere with the target chemical), or physical (materials that change the physical form of the target chemical by precipitation, adsorption, etc.).

JSLSCAD: Joint Service Lightweight Standoff Chemical Agent Detector.

Lidar: light detection and ranging—an active detection technique using a laser-like beam as the transmitter.

M21 RSCAAL: Remote Sensing Chemical Agent Alarm.
MCAD: Man-portable Chemical Agent Detector.
Mie scattering: scattering of radiation by solid particles or liquid droplets the diameters of which are approximately equal to the wavelength of the radiation.

Neural network: a calibration algorithm based on the concept of connectivity of modes. By constructing all possible connections between nodes, all possible relationships between input data and the final desired results can be explored.
NRC: National Research Council.

ORD: Operational Requirements Document.

Passive detector: detectors that require no action on the part of the operator to generate the analytical signal.

Simulant: a molecule having either spectroscopic or physical properties similar to the corresponding properties of a CWA.